SUPERVISED MACHINE LEARNING IN WIND FORECASTING AND RAMP EVENT PREDICTION

Wind Energy Engineering

SUPERVISED MACHINE LEARNING IN WIND FORECASTING AND RAMP EVENT PREDICTION

HARSH S. DHIMAN
Department of Electrical Engineering
Institute of Infrastructure Technology Research and Management
Ahmedabad, India

DIPANKAR DEB
Department of Electrical Engineering
Institute of Infrastructure Technology Research and Management
Ahmedabad, India

VALENTINA EMILIA BALAS
Department of Automation and Applied Informatics
Aurel Vlaicu University of Arad
Arad, Romania

ACADEMIC PRESS
An imprint of Elsevier

Academic Press is an imprint of Elsevier
125 London Wall, London EC2Y 5AS, United Kingdom
525 B Street, Suite 1650, San Diego, CA 92101, United States
50 Hampshire Street, 5th Floor, Cambridge, MA 02139, United States
The Boulevard, Langford Lane, Kidlington, Oxford OX5 1GB, United Kingdom

Notices

Knowledge and best practice in this field are constantly changing. As new research and experience broaden our understanding, changes in research methods, professional practices, or medical treatment may become necessary.

Practitioners and researchers must always rely on their own experience and knowledge in evaluating and using any information, methods, compounds, or experiments described herein. In using such information or methods they should be mindful of their own safety and the safety of others, including parties for whom they have a professional responsibility.

To the fullest extent of the law, neither the Publisher nor the authors, contributors, or editors, assume any liability for any injury and/or damage to persons or property as a matter of products liability, negligence or otherwise, or from any use or operation of any methods, products, instructions, or ideas contained in the material herein.

Library of Congress Cataloging-in-Publication Data
A catalog record for this book is available from the Library of Congress

British Library Cataloguing-in-Publication Data
A catalogue record for this book is available from the British Library

ISBN: 978-0-12-821353-7

For information on all Academic Press publications
visit our website at https://www.elsevier.com/books-and-journals

Publisher: Candice Janco
Acquisition Editor: Lisa Reading
Editorial Project Manager: Leticia M. Lima
Production Project Manager: Anitha Sivaraj
Designer: Victoria Pearson

Typeset by VTeX

Working together
to grow libraries in
developing countries

www.elsevier.com • www.bookaid.org

Dedicated to almighty, my parents Renu and Sanjay, brother Hardik, and the love of my life Shiwangi.

Harsh S. Dhiman

Dedicated to my wife Indulekha and my son Rishabh for providing me unending love and support.

Dipankar Deb

Dedicated to my husband Marius for patience, love, friendship, and humor.

Valentina Emilia Balas

Contents

List of figures

List of tables

Biography

About the series editor

Dr. Yves Gagnon is professor of engineering at the Université de Moncton, Canada, along with being Adjunct Professor at Thaksin University, Thailand. Prior to that, Dr. Gagnon was Visiting Executive at the Natural Sciences and Engineering Research Council (NSERC) of Canada, he was founding President and CEO of the New Brunswick Innovation Foundation and he held the positions of Associate Vice-President of Research, Dean of the Faculty of Graduate Studies and Research, and K.C. Irving Chair in Sustainable Development at the Université de Moncton.

As a Mechanical Engineer, Dr. Gagnon obtained graduate degrees from the Massachusetts Institute of Technology in the United States, and from the Université de Toulouse, in France, while Thaksin University has conferred him an Honorary Doctorate degree in Sciences. Active in research, Dr. Gagnon has been supported through various grants and research contracts. His research and innovation work, done in national and international collaborations, has appeared in over 350 publications and reports; and his contributions have been recognized by several prizes. Amongst his editorial positions, Dr. Gagnon was Associate Editor of the ASME Journal of Solar Energy Engineering, responsible of the wind energy section.

About the authors

Harsh S. Dhiman was born in Chandigarh in 1992 and is currently pursing PhD in Department of Electrical Engineering from Institute of Infrastructure Technology Research and Management (IITRAM), Ahmedabad, India, under the supervision of Prof. Dipankar Deb. He obtained his Master's degree in Electrical Power Engineering from Faculty of Technology & Engineering, Maharaja Sayajirao University of Baroda, Vadodara, India, and B. Tech in Electrical Engineering from Institute of Technology, Nirma University, Ahmedabad, India. His current research interests include hybrid operation of wind farms, hybrid wind forecasting techniques, and wake management in wind farms.

Dipankar Deb completed his PhD from University of Virginia, Charlottesville, under the supervision of Prof. Gang Tao, IEEE Fellow and

Professor in the department of ECE in 2007. In 2017, he was elected to be IEEE Senior Member. He has served as Lead Engineer at GE Global Research Bengaluru (2012–2015) and as Assistant Professor in EE, IIT Guwahati, 2010–2012. Presently, he is Professor and Department Coordinator of Electrical Engineering at Institute of Infrastructure Technology Research and Management (IITRAM), Ahmedabad. He is Associate Editor of IEEE Access Journal. He has authored 21 SCI Indexed Journal Papers, holds 6 US Granted patents, and has authored or edited eight books. His research interests include control theory, stability analysis, and renewable energy systems.

Valentina Emilia Balas holds PhD in Applied Electronics and Telecommunications from Polytechnic University of Timisoara, Romania. Dr. Balas is the author of more than 300 research papers in refereed journals and international conferences. Her research interests are in intelligent systems, fuzzy control, soft computing, smart sensors, information fusion, and modeling and simulation. She is the Editor-in Chief of International Journal of Advanced Intelligence Paradigms (IJAIP) and International Journal of Computational Systems Engineering (IJCSysE), an Editorial Board member of several national and international journals, and an evaluator expert for national and international projects and PhD theses. Dr. Balas is the director of Intelligent Systems Research Centre in Aurel Vlaicu University of Arad, Romania, and Director of the Department of International Relations, Programs and Projects at the same university. She served as General Chair of the International Workshop Soft Computing and Applications (SOFA) in eight editions 2005–2018 held in Romania and Hungary.

Dr. Balas participated in many international conferences as Organizer, Honorary Chair, Session Chair, and a member in Steering, Advisory, or International Program Committees. She is a member of SIAM, Senior Member of IEEE, a member in TC-Fuzzy systems (IEEE CIS), a member in TC-Emergent Technologies (IEEE CIS), a member in TC-Soft computing (IEEE SMCS). Dr. Balas was past Vice-President (Awards) of International Fuzzy Systems Association Council (2013–2015) and is Joint Secretary of the Governing Council of Forum for Interdisciplinary Mathematics (FIM), A Multidisciplinary Academic Body, India.

Preface

Recently, alternative energy sources have gained much importance owing to their clean operation. With limited fossil fuel and nuclear resources, solar and wind energy technologies have outgrown their market share. Given its rich sustainability, a renewable energy power portfolio strengthens the backbone of a country's economy. Apart from its positive environmental impact, wind energy has also globally created job opportunities. With increased penetration of renewable energy sources, their operation and control have become important. Wind turbines undergo a wide range of dynamic phenomena. Achieving economies of scale is the primary objective for a wind farm operation.

Given the random nature of wind speed, accurate wind forecasting schemes aid the operator to minimize the losses. Selecting an optimum operational strategy lowers the system cost and increases the reliability of the power system. Wind energy globally has impacted thousands of livelihoods in terms of employment opportunities and renewables-based power supply. 21st century has seen a spurt in wind energy installations worldwide, particularly, in offshore areas. However, the intermittency in wind power generation brings additional challenges in grid reliability. The stochastic wind nature of wind speed triggers the need for accurate wind forecasts ahead of time in order to plan optimal power dispatch. Traditional models based on numerical weather prediction and time-series based statistical models (ARMA, ARIMA, and S-ARIMA) face numerous shortcomings due to nonlinearity in wind speed.

The book deals with issues faced in short-term wind forecasting techniques and addresses hybrid machine intelligent algorithms that solve the issues with aforementioned traditional methods. Short-term wind speed forecasting based on signal decomposition techniques coupled with machine learning algorithm like support vector regression and its variants gives an accurate wind forecast. Variants of SVR like least square support vector regression (LSSVR), twin support vector regression (TSVR), and ε-twin support vector regression (ε-TSVR) are discussed.

Chapter 1 discusses the need for wind energy globally and detailed statistics related to installed capacity and estimated power production in coming years. It also throws light on the motivation and the objective behind this book. Chapter 2 highlights the fundamental concepts for wind energy starting from power production, wind speed distribution across on-shore and offshore regimes, and commonly used wind speed profile laws with changes in terrain.

Chapter 3 entails the paradigms of wind forecasting that focuses on forecasting needs, issues, and challenges. We also discuss the traditional forecasting schemes, and we progressively highlight the need for machine learning algorithms in short-term wind speed forecasting. Moving forward, Chapter 4 provides a deep insight into supervised machine learning algorithms specifically on support vector regression and its variants. In Chapter 5, random forest regression and gradient boosted machines (GBM) are presented with relevant literature and two case studies for rainfall and crude oil price prediction. We also discuss the importance of each method from the computation speed point of view.

Chapter 6 presents the hybrid model for wind forecasting based on wavelet transform and variants of support vector regression like ε-SVR, LSSVR, TSVR, ε-TSVR, RFR, and GBM. Case studies from wind farms are globally presented, and their results are discussed. Further, the effect of SVR hyperparameters on the regression quality is assessed. Chapter 7 discusses ramp events in wind farms, an analytical relationship between forecasting models is derived, and ramp-up and ramp-down events for different wind datasets are assessed based on their forecasting performance.

Chapter 8 discusses a specific application of wind speed prediction under the presence of wind wakes for supervised machine learning regression model based on conventional support vector regression. Case studies incorporating two different wind farm layouts are discussed.

Harsh S. Dhiman
Dipankar Deb
Valentina E. Balas

Acknowledgments

The achievement of this goal would not have been possible without assistance of Institute of Infrastructure Technology Research and Management (IITRAM), Ahmedabad, India, for necessary infrastructural support for writing this book. The authors are also grateful to Elsevier for allowing us to use related content for the book. We would like to thank the scientific and research organizations and agencies like Wind Energy Center, University of Massachusetts, for online access to wind speed datasets.

Acronyms

ACF	Autocorrelation function
ANN	Artificial neural network
ADM	Actuator disk model
ARMA	Autoregressive moving average
ARIMA	Autoregressive integrated moving average
DWT	Discrete wavelet transform
EMD	Empirical mode decomposition
ELM	Extreme learning machine
GA	Genetic algorithm
GBM	Gradient boosted machines
GRA	Grey relational analysis
GWEC	Global Wind Energy Council
HAWT	Horizontal axis wind turbine
KKT	Karush–Kuhn–Tucker
LSSVR	Least square support vector regression
ML	Machine learning
MAE	Mean absolute error
MAPE	Mean absolute percentage error
MLR	Multiple linear regression
MSE	Mean squared error
NMSE	Normalized mean squared error
SVR	Support vector regression
PDF	Probability density function
PSO	Particle swarm optimization
RFR	Random forest regression
RBF	Radial basis function
RMSE	Root mean squared error
SPSS	Statistical package for social sciences
TSVR	Twin support vector regression
ε-TSVR	ε-twin support vector regression

CHAPTER 1

Introduction

The demand for energy and the meteoric rise in greenhouse gas emissions due to the use of fossil fuels have opened doors for large investment in renewable energy. The production of electrical power based on wind power using wind turbines has become one of the key renewable sources since it can produce a clean and reliable energy with substantially low cost of production. In this introductory chapter, we focus on providing a general outline in the thrust area of machine learning application to wind forecasting and allied areas. The need for machine learning boosts the investment opportunities, which otherwise are limited due to false paradigms.

1.1 Renewable energy focus

The evolution of renewable energy technologies over the past decade has transcended all expectations. Global installed capacity and production from all renewable technologies have increased substantially, and supporting policies have continued to spread to more countries in all the regions of the world. It was during the early 1970s when the energy crisis and economic meltdown resulted in growth of renewable energy sources. Developing countries like Denmark, United States, Spain, and Germany initiated the renewable energy business owing to climate change. A particular energy source is regarded as "*sustainable*" if it preserves the social, economic, and environmental traits for the future generations. A renewable energy source is the one whose tangible viability remains unlimited and can be utilized in its raw form.

With coal reserves being limited, alternative sources of energy have entered the market to provide a sustainable and economic route to consume clean energy. The energy production from renewable sources is increasing at a steady rate and has led to economic growth in many developing countries. One of the primary reasons for promoting the renewable energy sources is the environmental concern causing depletion of ozone layer in atmosphere and greenhouse gas emissions. As a result of incessant industrial activities, the first international treaty, Kyoto Protocol, an extension to United Nations Framework Convention on Climate Change (UNFCCC) came into existence with a motto to curb artificial gas emissions [1]. These gas emis-

Supervised Machine Learning in Wind Forecasting and Ramp Event Prediction
https://doi.org/10.1016/B978-0-12-821353-7.00012-0
1

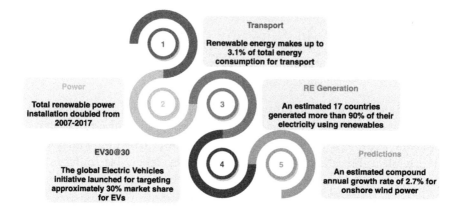

Figure 1.1 Top developments in renewable energy sector over past decade.

sions include carbon dioxide (CO_2), nitrous oxide (N_2O), methane (CH_4), and sulphur hexafluoride (SF_6). Further, the parties have committed to a nation–wide reduction of the emissions. However, there exist some countries that do not commit to this target due to their low per capita income. Some exceptions to the emissions in the form of international shipping and aviation also exist. The cleaner form of power production from renewable sources continues to bloom the energy sector, thus leading to huge investments from public and private sectors. In terms of the recent developments in renewable energy, Fig. 1.1 illustrates the top areas where the impact is more pronounced.

The installation count from renewable sources has doubled since 2007. Among renewable sources, wind and solar energies have been a focus for the policymakers sense inception of wind as a clean fuel. 2015 witnessed a 44% increase in wind power compared to 2014. With appropriate technology available, the power from renewable energy sources can be suitably tapped. Hydro, solar, and wind power technologies were seen as leading market drivers. With offshore wind power becoming more dominant in competitive market, a steady decline has been seen in the levelized cost of energy (LCOE). The prices of the renewable energy bids have been remarkably low (up to 30 USD per MWh) in European countries. China accounts for world's largest power production from renewable energy sources, a staggering 647 GW of installed capacity, which consists of 313 GW of power from hydro power sources. Countries like Germany, Sweden, and Denmark are focused on integrating increasingly larger shares of solar PV and wind power into utility grid systems by improving regu-

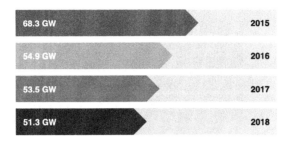

Figure 1.2 Global newly added wind power installations.

latory policies and by incorporating improved transmission system so as to broaden balancing areas. Meanwhile, countries are also heavily investing in energy storage for reducing a significant impact of intermittency.

1.2 Wind energy: issues and challenges

As discussed earlier, renewable energy sources have had significant impact on developing economies. Specifically, wind power has been in limelight owing to its clean and ecofriendly nature. Fig. 1.2 shows the popularity of wind power in terms of the installations since 2015.

Available in free form, the wind regime for a particular terrain essentially depends on weather variables like ambient air temperature, pressure, and humidity. However, wind power suffers from challenges like grid integration, power quality, reserve management, congestion management, and accurate forecasting.

Grid integration studies for wind power have revealed many technical and sociocultural challenges. The primary factor that determines a neat wind power grid integration is the generation capability of wind farms. For onshore wind farms, the proximity to utility grid has proved to be a beneficial factor that ultimately reduces the transmission and auxiliary costs. A controlled generation schedule for a wind farm results from accurate and reliable forecasts. In terms of flexibility, wind power faces enormous challenges like sudden change in wind speed magnitude and direction. This phenomenon is called a ramp event. Such events can be better handled if the wind power forecasts are made with high precision. Wind power forecasting started in the early 1980s to model wind speed as time series [2]. Brown et al. have studied the time-series models essentially to simulate the wind speed and power. Wind speed forecasts are transformed into wind power forecasts using a transformation law. First, the wind speed is fore-

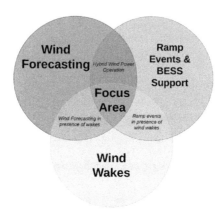

Figure 1.3 Core research areas in wind energy.

casted at a base height and then extrapolated to turbine hub height by using either logarithmic law or power law. Time-series models like autoregressive moving average (ARMA) and autoregressive integrated moving average (ARIMA) models have been applied to forecast wind speed and plan wind power dispatch for optimal power system operation. However, the accuracy obtained using these models is low due to various limitations posed by the nonlinear and stochastic nature of wind speed.

In terms of the industry preference, many wind forecasting tools based on prediction horizon have been developed. Wind forecasting stages can be categorized based on very short-term, short-term, medium-term, long-term, and very long-term forecasting. Short-term forecasting is a preferred choice owing to market operations and optimal power dispatch. These days electricity markets often follow deregulated structure allowing several competing bidding agents to participate in the retail process. Since power produced from wind resource is stochastic in nature, a day-ahead electricity market is often considered to plan the optimal dispatch. This dispatch is essentially dependent on the wind power schedules provided by the market operator. A stochastic wind power production brings errors in large forecast, which calls for the need for reserve capacity to compensate the deficit powers. If sufficient reserves are not available, then the cost incurred to the market operator escalates drastically. Since wind power penetration is increasing significantly daily, more efficient forecast schemes are required to ensure balance in the electricity markets.

Fig. 1.3 illustrates the core areas where the research in wind energy is concentrated. Wind forecasting, wind wakes, ramp events, and battery energy storage system (BESS) integration are important from the wind farm operator point of view.

As described earlier, wind forecasting governs the modern day electricity markets. However, within a wind farm, there are some inherent aerodynamic phenomena that reduce the overall efficiency. One such phenomenon is wind wakes. Wind wakes reduce the power capturing capacity of the downstream wind turbines in a wind farm due to drop in velocity, which can be essentially corrected to some extent by optimally placing wind turbines in a given area. The process of optimal wind turbine placement is called micrositing.

Power losses incurred due to improper wind turbine placement can also raise alarming situations of increased turbulence, which further can escalate the blade and tower health issues. Several optimization techniques like the genetic algorithm and particle swarm optimization are used to tackle such problems for an optimal wind farm layout, which are discussed in subsequent chapters ahead. Allied areas of wind forecasting also include the lifetime prediction of wind turbine blades and tower. Since commonly used turbine configuration is HAWT, the design is prone to extreme damages from high wind speed scenarios, called ramp events. A typical blade health management study includes prognostics and diagnostics, which are aimed to identify potential causes for poor blade design, remaining lifetime, and mechanical faults, which result in untimely degradation of blades. Studies have been carried out that proclaim deployment of sensors aiding online health monitoring.

Furthermore, ramp events, which are wind conditions of sudden change in speed and direction, are considered to be an alarming situation. These scenarios occur when the installed capacity is large incorporating turbines of high rated power. Ramp events are related not only to wind farm but also to many other scientific and engineering activities. In case of a reservoir wall, forest fires, air quality, and rapid changes in wind speed and direction can significantly impact these phenomena. We will further discuss these events in detail in Chapter 7. Overall, a wind farm operation is a sensitive one, which can be perceived from important aspects as discussed.

1.3 Machine learning in allied areas of wind energy

As discussed earlier, the study on wind forecasting has grown over the years with researchers focusing on improved prediction techniques. Forecasting wind speed stabilizes the modern day market in terms of the power dispatch and leads to a sensible planning of reserve power capacity. Since BESS have been actively used as a reserve in modern wind power plants, excessive charging and discharging of these battery units hamper their lifetime. A typical BESS capacity is determined from the charging and discharging schedules available from forecasted wind powers. Thus an accurate wind forecasting scheme can guarantee a wind farm operator to determine the BESS capacity for reserve power mode.

Since its inception in the 1970s, deep learning has been actively used in applications like regression, classification, pattern recognition, and clustering. Techniques like artificial neural networks (ANNs) have been put into use for predicting weather variables like solar radiation and wind speed. Since then, artificial intelligence that makes use of historical information has been used to improve the overall performance by using additional data. For wind farms, the source of data include SCADA, CMS data, maintenance data, and failure history. Software tycoons like Google have come up with their own forecast tool DeepMind, which is based on a neural network. The system is used to predict wind power for 36 hours ahead and has been applied successfully in the United States. Some of the commonly used applications of artificial intelligence (AI) are illustrated in Fig. 1.4.

Varied wind conditions impose fluctuations on the rotor blades and tower of wind turbines, which ultimately result in failure of the turbine. In case of fault detection, machine learning-based techniques are applied extensively to continuously monitor the wind turbine performance. Since wind turbines are made up of equipments of large pieces, wear and tear can cause damages at an enormous scale. Parameters like rotor speed, blade pitch angle, and yaw angle, which affect the turbine power capture, are often analyzed for the same. Commonly used AI methods include feature-based algorithms that essentially use the vital information catered by sensors mounted on turbine equipments. Variables like acoustic noise, temperature, electrical torque, rotor speed, and oil quality are often monitored continuously through sensors.

1.4 Scope and outline of the book

This book outlines the major paradigms related to wind forecasting and associated issues that are dealt with a daily basis. The wind power is directly proportional to the cube of the wind speed and hence possesses vital information regarding its variability, and therefore by forecasting wind speed we may save a lot of human hours for operation and maintenance (O&M).

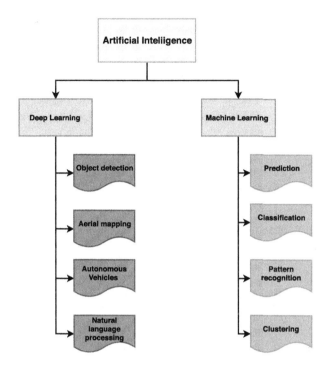

Figure 1.4 Common applications of artificial intelligence.

Wind forecasting is an important procedure, which directly impacts the reliability of utility grid. To address these issues, advanced forecasting algorithms like support vector regression (SVR), random forest regression (RFR), and gradient boosted machines (GBMs) are discussed in depth in this book.

The first part of this book deals with wind energy fundamentals in terms of wind power, major turbine components, leading manufacturers, wind speed distribution models, and turbine micrositing. Further, we discuss the thrust areas of wind forecasting and its modeling as a time series. Important statistical properties that test the stationarity of the time series are discussed along with R codes in Appendix A. Furthermore, a lucid and in-depth discussion on machine learning and its application to wind energy sector is presented. We also discuss the comparison between supervised and unsupervised learning and why supervised learning holds major relevance to wind forecasting.

The second part of this book addresses case studies from various wind farm sites for wind forecasting and ramp events. A hybrid forecasting method utilizing wavelet transform and empirical mode decomposition is discussed throughout and is applied for short-term wind speed forecasting. Machine learning-based regressors like SVR, RFR, and GBM are tested are their performance is evaluated using standard error metrics. Further ramp events are discussed specifically for onshore and offshore wind farm sites, and the best possible regressor is identified for predicting ramp events.

References

[1] UNFCCC, Kyoto Protocol to the United Nations Framework Convention on Climate Change, https://unfccc.int/resource/docs/convkp/kpeng.pdf, 2019. (Accessed 26 June 2019).
[2] B.G. Brown, R.W. Katz, A.H. Murphy, Time series models to simulate and forecast wind speed and wind power, Journal of Climate and Applied Meteorology 23 (8) (1984) 1184–1195.

CHAPTER 2

Wind energy fundamentals

In the past few decades the investments in wind energy sector have increased drastically. An abundant wind resource has led to increase in installed capacity worldwide. Energy demand pushes the envelope for renewable energy technologies, and solar, wind, and biomass being the pioneers, many developing countries have begun utilizing the sustainable sources of energy. Carbon footprints, which are created by stream power plants, can be significantly improved by utilizing power from wind resource. Operations and maintenance (O&M) sector has boomed with job opportunities with advent in wind power technology, thus decreasing the unemployment ratio in developing countries. According to Global wind energy council (GWEC) report [1], in 2017, with an installed capacity of 2.08 GW, the wind sector in South Africa created 15,000 jobs, 262,712 jobs in Europe, and 60,000 jobs in Indian subcontinent. Lucrative tariff rates have ensured support for wind technology, both onshore and offshore. Despite numerous benefits, the wind sector causes imbalance in aquatic life, high initial investment costs, and procedural issues in land acquisition. However, advanced manufacturing technologies cause rapid wind energy installations, and wind regime for offshore sites is much stronger than for onshore ones, thereby motivating investors to participate in the bidding process. Threats posed by wind turbines include bird killings, high noise levels, and opposition from local communities concerned about their livestock.

Wind farms are setup considering the space constraints, and recently much impetus has been given to Savonius-style wind turbines (SSWT) operating under any wind direction. Roy et al. [2] explored an inverse method based on differential evolution for computing optimal turbine parameters. Results reveal that area of SSWT is reduced by 9.8%. Further, a 2D CFD model is presented by Gupta and Biswas [3] to study the steady-state performance of a twisted three-bladed H-Darrieus rotor. Considering the stochasticity in wind speed, precise prediction of wind speed yields benefits to the wind farm operators. However, the error processing of forecasted wind speed/power and actual wind speed/power enables an appropriate choice of forecasting algorithms. Machine learning models like artificial neural networks (ANNs), support vector regression (SVR) [4,5], Gaussian

Supervised Machine Learning in Wind Forecasting and Ramp Event Prediction
https://doi.org/10.1016/B978-0-12-821353-7.00013-2

9

process regression (GPR), fuzzy logic, and extreme learning machine are in widespread usage.

2.1 Basics of wind power

Extracting power from moving air is not new as it has been practiced since ages. Growing energy demands have caused a positive push for renewable power sector. Among the renewables, wind energy is a promising area to provide uninterrupted power to the utility grid. Wind power is not only clean but also poses economic solutions to countries with wind resource. Wind farms are equipped with turbines that utilize the kinetic energy of the wind to generate usable power. This is further fed to utility grid via transmission lines and cables. Fundamentally, power extracted by a turbine having the rotor area A_r from the moving air with density ρ_w and velocity u_∞ is given as

$$P_w = \frac{1}{2}\rho_w A_r u_\infty^3. \tag{2.1}$$

Essentially, the power extracted is directly proportional to the wind speed u_∞ and rotor area A_r. However, with land constraint and manufacturing cost incurred, the rotor size poses a limitation in large–sized wind turbines. Thus regions with high wind speed can be suitably exploited to enhance the power capture capacity of wind farms. Further, (2.1) describes the theoretical power that can be potentially extracted from wind. However, in reality, only a fraction of this power is made available for mechanical-to-electrical conversion, and this fraction is known as the power coefficient. Wind turbines with different rotor configurations have different power coefficients. Theoretically, the maximum value of the power coefficient can be calculated based on the actuator disk model (ADM) of wind turbine, which assumes wind turbine as a disc that exerts forces on the fluid flow. The maximum value of the power coefficient is 0.5926 and is known as the Betz limit [6].

2.2 Wind resource assessment

Currently, wind installations have a total of 591 GW capacity with 51.3 GW newly added in 2018, out of which 4.5 GW is contributed by offshore wind farms as stated by the annual report of Global Wind Energy Council [7]. Recent developments in wind energy sector reveal that returns to the investment in building large wind parks is highly dependent on the nature of wind regime. While finalizing the site for wind farms, a special focus is laid on the wind turbine topology, wind speed distribution, wind forecasting, and nature of terrain. One of the key drivers of wind energy potential is the turbine technology put into use. Typically, two types of wind turbine configurations are used based on their axis of rotation, that is, horizontal-axis wind turbine (HAWT) and vertical-axis wind turbine (VAWT). Commonly used HAWTs find their application in commercially small- and large-scale electric power generation, whereas the VAWTs are preferred mostly for small-scale electric power generation, that is, less than 100 kW. Renowned companies like General Electric, Gamesa, Vestas, Siemens, and Suzlon group have been actively involved in manufacturing market of wind turbines.

Wind farm companies whose main objective is profit maximization lay their focus on wind resource assessment through optimal usage of wind resources. Scientific analysis of wind regime aimed to determine the true potential of a wind site quantitatively and qualitatively in terms of wind speed variability, turbulence characteristics, and nature and frequency of ramp events. This book aims to provide insights into different aspects that involve these attributes so as to understand ways and means of maximizing power output from wind farm. Even after the wind farm has been commissioned and is operating, it is still possible to install extensions to rotor blades, for instance, so as to improve productivity. Effective blade health monitoring is another wind resource assessment that can be very effective in long run.

2.3 Wind speed distribution

The wind speed distribution of a particular regime depends on the topographical features like air temperature, pressure, humidity, and so on. Various wind speed distribution functions like Weibull, Rayleigh, lognormal, and gamma distributions have been proposed in the literature [8]. Amongst the listed probability distribution functions, the Weibull distribution is most commonly used because of its simplicity.

The Weibull distribution is widely used in applications like reliability analysis (failure rate), industrial engineering (to represent manufacturing and delivery time), and wind speed distribution [9]. The Weibull distribution holds several advantages like flexibility and generally gives a good fit to the observed wind speed [10]. However, the Weibull distribution cannot accurately estimate all the wind regimes including that of bimodal wind

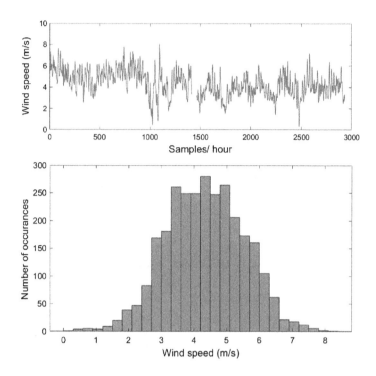

Figure 2.1 Frequency distribution of wind speed data, Praia Formosa, Brazil, 2017.

distributions involving two peaks. Various other distributions like Rayleigh and log-normal distributions have also been quoted in the literature. Studies have shown that the two-component Weibull distribution works well with bimodal wind speed distributions. Similar to the Weibull distribution, mixture density functions have also been proposed [11]. Fig. 2.1 illustrates the frequency distribution plot for a wind speed data of Praia Formosa wind farm in Brazil.

From the figure we can observe that the frequency distribution has two peaks of wind speed thereby indicating a bimodal distribution. The wind speed data is for a duration of four months from January 1, 2017, to April 30, 2017. The frequency distribution plot helps to investigate the statistical properties of the wind regime in terms of maximum, minimum, and mean wind speed.

2.3.1 Probability density functions for wind speed

Probability density functions are widely used in statistics and in other computational fields, where the probability of occurrence of random variable such as wind speed is described mathematically. However, the wind speed is not the only variable that is affecting the power captured by the wind turbine. The correlation between wind speed and wind direction is equally important. The literature also shows implementation of mixture density functions, which are composite probability density functions, each with an associated weight [12].

2.3.1.1 Weibull distribution

The majorly used wind PDF is Weibull, which fits well for almost all wind regimes. The two-parameter (k and λ) Weibull PDF is represented as

$$f(v; k, \lambda) = \frac{k}{\lambda}\left(\frac{v}{\lambda}\right)^{k-1} e^{-(\frac{v}{\lambda})^k}, \tag{2.2}$$

where $k, \lambda > 0$, and $v > 0$ is a random variable called the wind speed (in m/sec); k is known as the shape parameter. Estimation of Weibull parameters can be done by the maximum likelihood method [13]. Similarly, joint Weibull PDFs can be used to model wind farms with bimodal wind regimes. A two-component joint Weibull PDF can be expressed as

$$f_i(v; k_i, \lambda_i) = \frac{k_i}{\lambda_i}\left(\frac{v}{\lambda_i}\right)^{k_i-1} e^{-(\frac{v}{\lambda_i})^{k_i}}, \quad i = 1, 2,$$

$$f(v; k, \lambda) = w_1 f_1(v; k_1, \lambda_1) + (1 - w_1) f_2(v; k_2, \lambda_2), \tag{2.3}$$

where k_1, λ_1 and k_2, λ_2 represent the shape and scale parameters for individual PDFs, which in turn can be estimated using maximum likelihood estimation (MLE).

Parameter estimation using maximum log likelihood

The estimates $\hat{\theta} = (\hat{k}, \hat{\lambda})$ of parameters of the Weibull distribution can be determined for a unique wind speed data based on maximum likelihood principle. Given the Weibull PDF as per Fig. 2.2 and (2.2), the log likelihood function is given as

$$L(v|k, \lambda) = n\log(k) - nk\log(\lambda) + (k - 1)\sum_{i=1}^{n}\log(v_i) - \sum_{i=1}^{n}\left(\frac{v_i}{\lambda}\right)^k. \tag{2.4}$$

Further, to obtain the estimates, the log likelihood function is differentiated with respect to parameters k and λ:

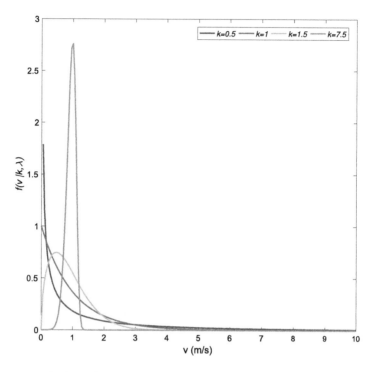

Figure 2.2 Weibull PDF for $\lambda = 1$ and different values of shape parameter.

$$\frac{\partial L(v|k,\lambda)}{\partial \lambda} = \frac{-n\lambda}{k} + \frac{\lambda}{k}\sum_{i=1}^{n}\left(\frac{v_i}{\lambda}\right)^k, \tag{2.5}$$

$$\frac{\partial L(v|k,\lambda)}{\partial k} = \frac{n}{k} - n\log(\lambda) + \sum_{i=1}^{n}x_i - \sum_{i=1}^{n}\left(\frac{x_i}{\lambda}\right)^k\log\left(\frac{x_i}{\lambda}\right). \tag{2.6}$$

Based on the solutions of (2.5) and (2.6), the parameters k and λ can be estimated. Looking at (2.4), the parameters can be obtained by minimizing the negative of the log likelihood function. Since this is an optimization problem, various evolutionary techniques, like the genetic algorithm (GA), particle swarm optimization (PSO), and firefly algorithm (FA), can be used to obtain accurate parameter estimates. A Weibull distribution fit is shown in Fig. 2.3.

2.3.1.2 Lindley distribution for wind speed

The unimodal wind regimes can be described by Weibull and gamma distributions [11]. With extremely low and high wind regions, Weibull and

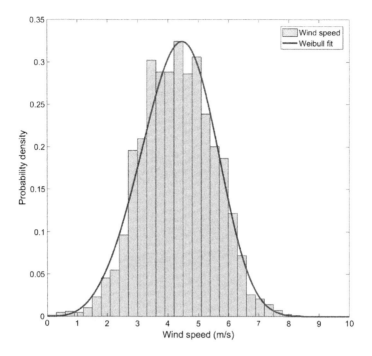

Figure 2.3 Weibull distribution fit for real-time wind speed data.

other PDFs are reported to not perform well, and in such cases the Lindley distribution function is suggested [14]. The Lindley distribution has two versions, namely, generalized Lindley and power Lindley, as described by [14]. The generalized Lindley f_{gl} and power Lindley f_{pl} PDFs can be represented for a random variable wind speed v as

$$f_{gl}(v) = \frac{\alpha\beta^2}{\beta+1}(1+v)\left(1 - \frac{1+\beta+\beta v^2}{1+\beta}e^{-\beta v}\right)^{\alpha-1}e^{-\beta v}, \qquad (2.7)$$

$$f_{pl}(v) = \frac{\alpha\beta^2}{\beta+1}(1+v^\alpha)v^{\alpha-1}e^{-\beta v^\alpha}, \qquad (2.8)$$

where α and β represent the shape and scale parameters, respectively.

The cumulative PDF for generalized Lindley distribution is expressed as

$$f_{cgl}(v) = \left[1 - \left(\frac{1+\beta+\beta v}{\beta+1}\right)e^{-\beta v}\right]^\alpha. \qquad (2.9)$$

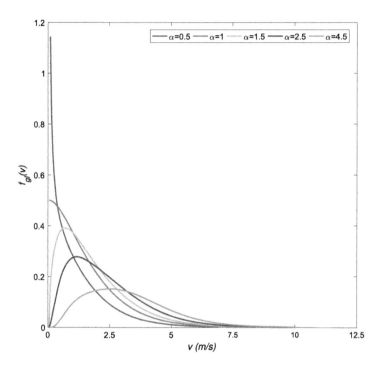

Figure 2.4 Generalized Lindley PDF for $\beta = 1$ and different values of α.

The extended generalized Lindley distribution for random variables and its superiority over well-known Weibull, Rayleigh, two-component Weibull, and generalized Lindley PDFs have been described in [15].

Fig. 2.4 illustrates the plots of the generalized Lindley PDF used to model the wind speed distribution. The shape and scale parameters α and β govern the peak and tail structures of the distribution. The generalized Lindley PDF can be particularly used to model the wind resource with weak regime in terms of wind speed. Small-scale wind farms with turbines having rated power in the range 100–500 kW operate at low wind speeds, for which the distribution is likely to be skewed.

Parameter estimation for Lindley PDF

The Lindley PDF is a two-parameter distribution class, where $\hat{\theta} = (\hat{\alpha}, \hat{\beta})$ are determined based on the log likelihood principle. The log likelihood function for Lindley PDF is given as

$$L(v|\alpha, \beta) = n[\log \alpha + 2\log \beta - \log(\beta + 1)] + \sum_{i=1}^{n} \log(1 + v_i)$$

$$+ (\alpha + 1) \sum_{i=1}^{n} \log \left(1 - \frac{1 + \beta + \beta v}{1 + \beta} e^{-\beta v}\right) - \beta \sum_{i=1}^{n} v_i. \qquad (2.10)$$

Further, differentiating the log likelihood function with respect to α and β, we get

$$\frac{\partial L(v|\alpha, \beta)}{\partial \alpha} = \frac{n}{\alpha} + \sum_{i=1}^{n} \log \left(1 - \frac{1 + \beta + \beta v}{1 + \beta} e^{-\beta v}\right), \qquad (2.11)$$

$$\frac{\partial L(v|\alpha, \beta)}{\partial \beta} = \frac{n}{\beta} - \sum_{i=1}^{n} x_i + (\alpha + 1) \sum_{i=1}^{n} \left[\frac{1 - \frac{e^{-\beta v}}{(1+\beta)^2} + v e^{-\beta v}}{1 - \frac{1+\beta+\beta v}{1+\beta} e^{-\beta v}}\right]$$

$$- (\alpha + 1) \sum_{i=1}^{n} \left[\frac{\frac{e^{-\beta v}}{(1+\beta)^2} + v e^{-\beta v}}{1 - \frac{1+\beta+\beta v}{1+\beta} e^{-\beta v}}\right]. \qquad (2.12)$$

Since Eqs. (2.11) and (2.12) have nonlinear terms, the solutions of α and β can be obtained using optimization methods like quasi-Newton methods that converge faster than the Newton method. The estimates $\hat{\alpha}$ and $\hat{\beta}$ are determined using the PSO algorithm. Fig. 2.5 illustrates the generalized Lindley fit for real-time wind speed data.

2.3.1.3 Stacy–Mihram distribution model

Another class of Weibull distribution model is the Stacy–Mihram model, which generally adequately describes the wind speed distribution. The two-parameter model is given as

$$f_{stm}(v) = \frac{\beta v^{\beta-1}}{\alpha^\beta} e^{-\left(\frac{v}{\alpha}\right)^\beta}, \quad v > 0. \qquad (2.13)$$

Parameters α and β can be estimated using MLE, for which the log likelihood function is given as

$$L(v|\alpha, \beta) = n\log \beta - n\beta \log \alpha + (\beta - 1) \sum_{i=1}^{n} \log(v_i) - \sum_{i=1}^{n} \left(\frac{v_i}{\alpha}\right)^\beta. \qquad (2.14)$$

Fig. 2.6 shows the Stacy–Mihram fit for a wind speed time-series.

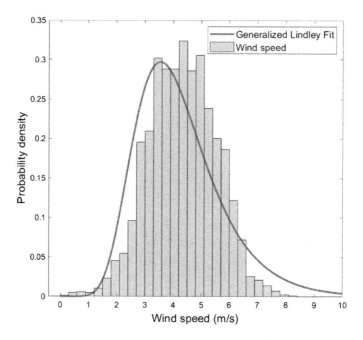

Figure 2.5 Generalized Lindley distribution fit for real-time wind speed data.

2.4 Wind turbine micrositing

Wind power extraction is dependent not only on the temporal distribution but also on the spatial distribution. As the number of wind turbines in a given wind farm increase, their lateral and longitudinal spacing affects the effective wind speed encountered at each turbine. Further, the terrain features for a particular wind farm site impact the power captured. Wind shear resulting from high wind speeds is also likely to cause damage to the blade geometry and tower strength. The process of optimal placement of wind turbines in a wind farm to maximize power capture and minimize cost of generating electricity is known as micrositing.

Early studies in micrositing have dealt with the genetic algorithm being put into use for optimal wind farm layout as described by Mosetti et al. [16], where a wind farm with an area of 4 km^2 is divided into cells of size $5D$, where D is the rotor diameter. Two cases, the first one with single wind speed direction and the second one with multiple wind direction, are considered. Results reveal that a random wind farm layout with 40 to 50 turbines has an efficiency of 50%, whereas an optimized wind

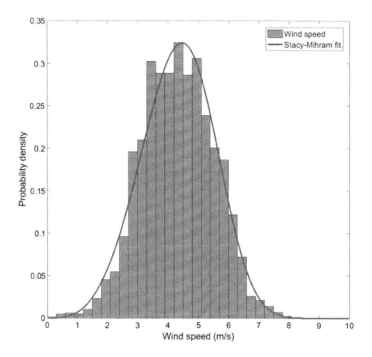

Figure 2.6 Stacy–Mihram distribution fit for real-time wind speed data.

farm layout has an efficiency of 95%. Further, Wan et al. [17] have carried out wind farm micrositing with PSO technique. The results of wind farm optimization are compared with study carried out by Grady et al. [18] and reveal a 6.34% improvement in wind power capture for a uniform wind direction with speed 12 m/sec. Wan et al. [19] have presented a Gaussian PSO (GPSO) algorithm with local search strategy to determine the optimal wind farm layout. To improve the micrositing, a local search strategy is applied to GPSO, and results reveal a 14.61% increase in wind farm efficiency when compared to empirical scheme. Multinational companies like General Electric have worked extensively in micrositing in last couple of decades to significantly improve the annual energy production (AEP) of a wind farm.

Fig. 2.7 illustrates the wind farm layouts with and without micro-siting. The first layout is a randomized one with maximum power losses due to wake interactions, whereas the second layout is an optimized one with an improved power capture capacity.

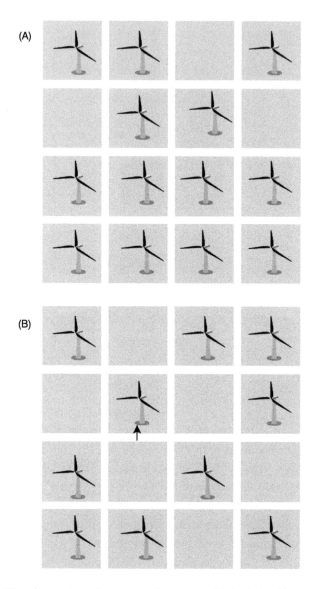

Figure 2.7 Wind farm micrositing. (A) Random layout. (B) Optimized layout.

References

[1] Global wind energy council report, http://files.gwec.net/files/GWR2017.pdf, 2017. (Accessed 7 March 2019).

[2] S. Roy, R. Das, U.K. Saha, An inverse method for optimization of geometric parameters of a Savonius-style wind turbine, Energy Conversion and Management 155 (2018) 116–127.

[3] R. Gupta, A. Biswas, Computational fluid dynamics analysis of a twisted three-bladed H-Darrieus rotor, Journal of Renewable and Sustainable Energy 2 (4) (2010) 043111.

[4] M. Mohandes, T. Halawani, S. Rehman, A.A. Hussain, Support vector machines for wind speed prediction, Renewable Energy 29 (6) (2004) 939–947.

[5] J. Zhou, J. Shi, G. Li, Fine tuning support vector machines for short-term wind speed forecasting, Energy Conversion and Management 52 (4) (2011) 1990–1998.

[6] S. Mathew, Wind Energy, Springer Berlin Heidelberg, 2006.

[7] GWEC conferences and seminars – GWEC, https://gwec.net/events/gwec-conferences-and-seminars, April 2019. (Accessed 31 May 2019).

[8] J. Carta, P. Ramírez, S. Velázquez, A review of wind speed probability distributions used in wind energy analysis, Renewable & Sustainable Energy Reviews 13 (5) (2009) 933–955.

[9] C.-D. Lai, D. Murthy, M. Xie, Weibull Distributions and Their Applications, Springer Handbook of Engineering Statistics, Springer London, 2006, pp. 63–78.

[10] T. Ouarda, C. Charron, J.-Y. Shin, P. Marpu, A. Al-Mandoos, M. Al-Tamimi, H. Ghedira, T.A. Hosary, Probability distributions of wind speed in the UAE, Energy Conversion and Management 93 (2015) 414–434.

[11] T.P. Chang, Estimation of wind energy potential using different probability density functions, Applied Energy 88 (5) (2011) 1848–1856.

[12] J. Wang, J. Hu, K. Ma, Wind speed probability distribution estimation and wind energy assessment, Renewable & Sustainable Energy Reviews 60 (2016) 881–899.

[13] J. Seguro, T. Lambert, Modern estimation of the parameters of the Weibull wind speed distribution for wind energy analysis, Journal of Wind Engineering and Industrial Aerodynamics 85 (1) (2000) 75–84.

[14] T. Arslan, S. Acitas, B. Senoglu, Generalized Lindley and power Lindley distributions for modeling the wind speed data, Energy Conversion and Management 152 (2017) 300–311.

[15] Y.M. Kantar, I. Usta, I. Arik, I. Yenilmez, Wind speed analysis using the extended generalized Lindley distribution, Renewable Energy 118 (2018) 1024–1030.

[16] G. Mosetti, C. Poloni, B. Diviacco, Optimization of wind turbine positioning in large windfarms by means of a genetic algorithm, Journal of Wind Engineering and Industrial Aerodynamics 51 (1) (1994) 105–116, https://doi.org/10.1016/0167-6105(94)90080-9.

[17] C. Wan, J. Wang, G. Yang, X. Zhang, Optimal Micro-Siting of Wind Farms by Particle Swarm Optimization, Lecture Notes in Computer Science, Springer Berlin Heidelberg, 2010, pp. 198–205.

[18] S. Grady, M. Hussaini, M. Abdullah, Placement of wind turbines using genetic algorithms, Renewable Energy 30 (2) (2005) 259–270.

[19] C. Wan, J. Wang, G. Yang, H. Gu, X. Zhang, Wind farm micro-siting by Gaussian particle swarm optimization with local search strategy, Renewable Energy 48 (2012) 276–286.

CHAPTER 3

Paradigms in wind forecasting

With increasing human demands for electricity consumption, the supply side management has taken a pivotal role in regulatory markets. Fossil fuel-based power generation is facing stern competition from renewable energy sources. The penetration of wind power, specifically, has increased owing to pollution free power generation and flexibility with transmission. The balance between demand and supply of power is an important criterion for power system regulators.

For optimal power dispatch, the generation companies have a reserve capacity in case the power flow from main generation entity ceases. However, the reserve capacity of power has some limits in terms of size and dispatchability. Therefore forecasting the load ahead of time can ascertain the operator to plan the power dispatch optimally.

3.1 Introduction to time series

Forecasting a particular variable can depend upon temporal or spatial scale. Temporal variations with time reflect the stochasticity present in the variable. Spatial variations usually are dominant in climatology and meteorology. A temporal scale for a variable can be modeled in terms of time series. A time series is a successively ordered sequence of numerical data points and can be taken on any variable changing with time.

A time-series-based analysis helps the operator to identify the dynamic behavior of the variable in terms of relative dependence of each time-step with respect to another. Fundamentally, a time series S_t can be defined as a set of observations for a particular variable s_t at each time-step t. As an example of time series, Fig. 3.1 illustrates the opening stock price of Facebook, Inc., from January 3, 2018, to June 3, 2019. Each data point in this time series indicates a unique value of stock price.

Essentially, time series can be categorized into stationary and nonstationary time-series. The statistical nature of a stationary time series in terms of mean and variance does not change over time. Mathematically, a stationary time series S_t can be expressed as

Supervised Machine Learning in Wind Forecasting and Ramp Event Prediction
https://doi.org/10.1016/B978-0-12-821353-7.00014-4

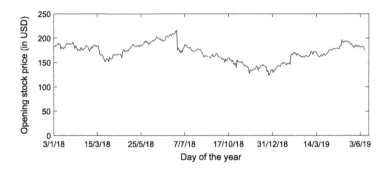

Figure 3.1 Time-series plot for opening stock price of Facebook, Inc.

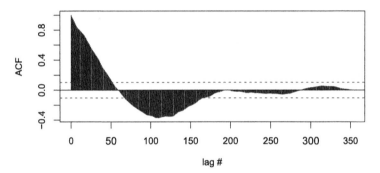

Figure 3.2 ACF plot for opening stock prices of Facebook, Inc.

$$F_S(s_{t_{1+h}}, \ldots, s_{t_{n+h}}) = F_S(s_{t_1}, \ldots, s_{t_n}), \tag{3.1}$$

where t is the time step, and h is the shifted time step.

To check whether the given time series is stationary or not, several statistical tests are available like the autocorrelation function (ACF), the augmented Dickey–Fuller (ADF) t-statistic test for unit root, the Ljung–Box test, and the Kwiatkowski–Phillips–Schmidt–Shin (KPSS) test for level or trend stationarity.

We now give two examples of R codes for checking the stationarity of a time series.

Example 3.1. R code for Autocorrelation Function

```
acf(VarName, lag.max=length(VarName), xlab="lag#", ylab='ACF',
main='~')
```

Autocorrelation plots reveal correlation of the elements of a variable with a time-shifted signal of itself. Fig. 3.2 depicts the ACF function for stock price time series. Qualitative analysis states that the ACF values exceed the confidence bounds thereby indicating nonstationarity of the time series.

Similarly, the Ljung–Box test and augmented Dickey–Fuller test can be carried out for determining whether the given time series is stationary or not.

Example 3.2. R code for the Ljung–Box test

```
lag.length=25 Box.test(VarName, lag=lag.length, type="Ljung-Box")
```

Output

```
data: VarName
X-squared = 5626, df = 25, p-value < 2.2e-16
```

3.2 Wind forecasting: overview

With the spurt in the installation of renewable energy sources, the fossil fuel usage is needed to be put into restricted use. Most of the industries in power sectors prefer renewable energy generation owing to its negligible carbon footprint. With wind available in abundant form, tapping power from wind is a specialized task. The error processing of forecasted wind speed/power and actual wind speed/power play a crucial role in selecting an appropriately applicable forecasting algorithm.

Wind forecasting plays an important role when it comes to clearing day ahead market scenarios. Given that there is a market situation to be cleared, an accurate wind forecasting scheme is helpful in such situations. Wind forecasting schemes are broadly categorized as (i) weather-based prediction methods and (ii) statistical or time-series-based prediction methods. When we consider weather-based prediction models, the wind forecast accuracy depends largely on the topology of the land where the wind turbines are erected.

Given the topology of the land, wind speed measurements at an appropriate height from the ground, the temperature of the ambient air, air pressure, and so on are important factors to be taken into consideration for improved forecasting. On the other hand, statistical methods solely depend on the past measurements of the wind speed to appropriately predict the future values.

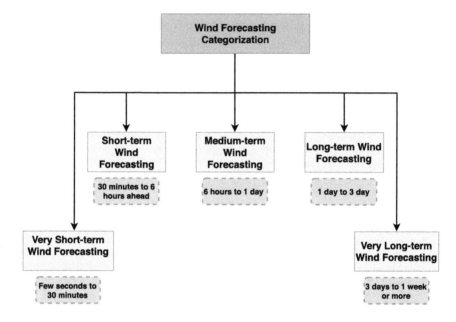

Figure 3.3 Categorization of forecasting methods based on prediction horizon.

3.2.1 Techniques based on prediction horizon

Wind speed forecasting applications majorly lie in the area of electricity market clearing, economic load dispatch and scheduling, and sometimes provide ancillary support. Thus a proper classification based on the prediction horizon, that is, the duration of prediction, becomes important for various transmission system operators. The time-scale classification of wind forecasting methods is given as shown in Fig. 3.3.

The classification not only simplifies the study but also helps to choose the accurate method depending on the type of its application. It appears that short-term and medium-term prediction methods are more in use due to their accuracy and robustness.

The time-scale classification proves to be a brownie point in terms of application of these forecasting algorithms. For any forecasting method to be accurate, its performance at different sites with varied atmospheric conditions plays a crucial role in forecasting future values. Thus a parameter must be assigned that not only calculates how accurate the method is but also determines its fitness over certain conditions. Such type of optimization studies are often done to choose the accurate forecasting method. Very short-term forecasting methods, as the name suggests, predict the future values up to a short span of time. Usual time-scale followed for this method is from a few seconds to 30 minutes. The core area of very-short-term wind forecasting lies in market clearing and regulatory actions. Since the global electricity market is now moving to deregulated markets, very-short-term and short-term forecasting has gained importance.

Of the many methods described in literature, a specific one like the spatial correlation method [1], where the 1-second ahead forecasting is done, and the artificial neural network (ANN)–Markov chain (MC) model [2], which uses a variable set of 175 min and predicts wind speeds for the next 7.5 s ahead. Bayesian structural break model [3] is used to predict 1 min and 1 h ahead forecasts. Data mining approach is also used to forecast wind speeds for a duration from 10 s to 5 min ahead [4].

Apart from these methods, several intelligent learning methods like artificial neural networks (ANNs) and, in particular, deep neural networks (DNNs) [5] are used for forecasting wind speeds up to 10 min, 30 min, and 1 h ahead. Support vector machines (SVMs), hybrid methods combining empirical wavelet transform (EWT) with neural networks (NNs) also form core of the very-short-term wind forecasting process. Recent advances in the forecasting domain show evolution of machine learning algorithms, with decomposition forecasting algorithms (DFA) being most in use. The essence of a decomposition algorithm lies in breaking the time series of a variable, that is, wind speed/power in the present case, and analyzing individual units for forecasting and combining them to obtain the resultant series.

One such decomposition algorithm is used in [6], where the time-series data of wind are broken or decomposed into several units, and on each unit a feature construction process is performed, and then those with best features are chosen for prediction. The prediction can be carried out with standard ARIMA model of order (p, d, q), or artificial neural networks (ANNs), or support vector machines (SVMs). Further, a forecasting technique based on the Hilbert–Huang transform (HHT), which uses the decomposition technique for nonstationary and nonlinear models, is also used [7].

Apart from these standard ARIMA models, a hybrid model, wind forecasting by wavelet transform and neural networks [8], is used to predict the wind power for a 3-h ahead horizon. Data from previous 12 h with 15-min time step is taken as input to the ANN input layer. The original wind power series is decomposed using the wavelet transform (specifically, the discrete wavelet transform (DWT)), and the resulting series is fed to the neural network where the future values are forecasted. The wavelet transform also finds use in forecasting the load and electricity prices for a power plant, particularly, in deregulation market [9]. Another short-term forecasting method involves the empirical mode decomposition (EMD) and feature selection [6], where the wind series is broken down into several subsequent series.

For each of these series, an intrinsic mode function (IMF) is computed. The IMF of a decomposed signal represents the irregularity and frequency components of the signal. Once the series is decomposed, the appropriate forecasting tool, ANN or SVM, is chosen to forecast the wind power values. In another method a combination of the wavelet transform, support vector machines, and the genetic algorithm is described [10]. The process of genetic algorithm helps to select the input parameters for the SVM. It is necessary to have optimized input for choosing the best candidates for the forecast. The results showed that the MAPE obtained with WT-SVM-GA is around 14.79%, and that with persistence is 22.64%.

Long-term forecasts find a major application in the field of unit commitment decisions, maintenance scheduling, and so on. A hybrid method [11] combining particle swarm optimization (PSO) and adaptive-network-based fuzzy interference system (ANFIS), that is, PSO+ANFIS, is used for forecasting 1-day ahead in the time steps of 15 min. The PSO algorithm is chosen to find out the best parameters for neuro-fuzzy systems. Another method [12] uses relevance vector machine (RVM), based on the wavelet decomposition (WD) and artificial bee colony optimization (ABCO), where a wind signal is decomposed into different subseries of frequency ranges, and then the forecasting is done by RVM. The kernel parameters of RVM are chosen by a metaheuristic algorithm known as the artificial bee colony optimization (ABCO).

Apart from the mentioned forecast techniques, many numeric weather methods such as global forecast system [13] and fifth generation mesoscale model (MM5) with neural networks are used. The predictions from the global forecast systems (GFS) plus the atmospheric conditions of the topography concerned are used as boundary conditions for the MM5 model to generate an output, which is then given as an input to the neural network. From the global forecast method, the physical downscaling of the wind data is done, and then using the MM5 model, the statistical downscaling is achieved.

3.2.2 Techniques based on methods of forecasting

Based on the methods of forecasting, several techniques are implemented for accurate wind speed forecasts. Traditionally, the persistence method is used to predict the future values of the variables under the assumption that the conditions remains unchanged between "present" time t and forecast time $t + h$. Mathematically, for a variable v, the persistence algorithm is given as

$$\hat{v}_{t+h} = v_t \tag{3.2}$$

and is best suited for variables with stationary time series as they possess identical statistical properties.

Moving forward, statistical models like ARMA, ARIMA, and f-ARIMA are also used to forecast wind speed for very-short-term and short-term-based forecasting. These models predict future values based on the past data. Machine learning-based methods are gaining popularity owing to their ability to handle large historical data. Methods like the support vector regression, artificial neural network, extreme learning machine, and Gaussian process regression are some of the commonly used machine learning techniques for wind forecasting. Since the wind speed is a stochastic time series, various signal decomposition techniques like the wavelet transform, empirical mode decomposition, ensemble empirical mode decomposition, and complementary ensemble empirical mode decomposition are used in tandem with machine learning methods to forecast the short-term wind speed.

Fig. 3.4 illustrates different forecasting techniques based on forecasting methods.

3.3 Statistical methods

In benign terms, statistical forecasting methods indicate the use of statistics based on the available historical quantitative data to ascertain what could happen in the future.

Short-term wind forecasting is the most commonly used forecasting category, which many day-ahead markets need to clear the market scenarios by the end of the day. Among these, the most used forecasting methods are a combination of two or more machine learning methods combined with time-series models like AR, MA, ARMA, ARIMA, and ARMAX. Auto regression (AR) is a time-series model that uses observations taken from previous time steps as input to a regression equation to forecast the

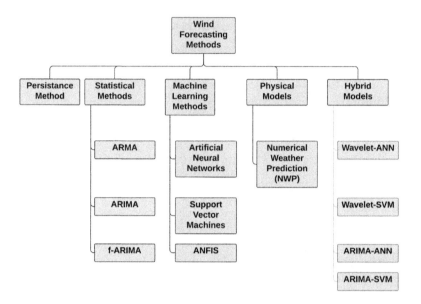

Figure 3.4 Categorization of wind forecasting techniques based on methods.

outcome at the next time step. On the other hand, rather than using past values of the forecast variable in a regression, a moving average (MA) model uses past forecast errors in a regression-like model.

Autoregressive exogenous moving average (ARMAX) is a nonlinear model that captures all the uncertainties related to the stochastic nature of the wind. Let us look at the two other models, ARMA and ARIMA, in more detail.

3.3.1 ARMA model

Among various statistical models, ARMA model is a popular forecasting method. In ARMA, AR stands for autoregressive, and MA stands for moving average. Thus a combined autoregressive (of order p) and moving average (of order q) forms an ARMA(p, q) model. The order of ARMA models, p for AR and q for MA, denotes the lag between present and past values of the variable under test. Thus a generic ARMA model can be mathematically expressed as

$$
(y_t - \mu_x) = \phi_1(y_{t-1} - \mu_x) + \phi_2(y_{t-2} - \mu_x) + \cdots + \phi_p(y_{t-p} - \mu_x)
$$
$$
+ \xi_t - \theta_1 \xi_{t-1} - \theta_2 \xi_{t-2} - \cdots - \theta_q \xi_{t-q}, \tag{3.3}
$$

Table 3.1 Error metrics for forecast accuracy.

Metric	Formula
R^2	$\frac{\sum_{i=1}^{n}(\hat{x}_i - \bar{x})^2}{\sum_{i=1}^{n}(x_i - \bar{x}_i)^2}$
RMSE	$\left(\frac{1}{n}\sum_{i=1}^{n}(\hat{x}_i - x_i)^2\right)^{1/2}$
MAE	$\frac{1}{n}\sum_{i=1}^{n}|\hat{x}_i - x_i|$
MAPE	$\frac{1}{n}\sum_{i=1}^{n}|\hat{x}_i - x_i| \times 100\%$

where ξ_t is an independent process with mean zero, and $\phi = [\phi_1, \ldots, \phi_p]$ and $\theta = [\theta_1, \ldots, \theta_q]$ are the parameters of the AR and MA processes, respectively. ARMA models can be studied in a much more simplified manner by using the lag operator L. When applied k times, the lag operator moves index of the variable back by k times. Given a quantity y, the lag operation for k times yields $L^k(y) = y_{t-k}$. In general, an ARMA process can be mathematically expressed as

$$\text{ARMA} : \phi L(y_t) = \theta L(\xi_t), \tag{3.4}$$

$$\text{AR} : \phi L(y_t) = \xi_t, \tag{3.5}$$

$$\text{MA} : y_t = \theta L(\xi_t). \tag{3.6}$$

As an example, we consider forecasting wind speed based on time-series modeling. The wind speed data of 2 months (Jan 1, 2017–March 31, 2017) for Praia Formosa, Brazil, is taken. The autocorrelation of the wind speed time series is calculated, and the orders p and q for ARMA model are set. To assess the model accuracy, error metrics like root mean squared error (RMSE), mean absolute error (MAE), mean absolute percentage error (MAPE), and R^2 are calculated and listed in Table 3.1.

The $\mathcal{L}_2$ norm or Euclidean distance of residuals is also used for investigating the forecast accuracy, which is expressed as

$$\| \mathcal{L} \|_2 = \sqrt{\sum_{i=1}^{n}(x_i - \hat{x}_i)^2}, \tag{3.7}$$

where x_i is the actual value, and $\hat{x}_i$ is the predicted value for n samples.

3.3.2 ARIMA model

ARMA models have a limitation that they can be applied to stationary processes only. In reality, the wind speed is a nonstationary time series with significant diurnal, meteorological, and seasonal variations. These variations are a result of dynamic atmospheric processes that change the wind behavior in time. To handle such limitations, ARIMA process is used to forecast wind speed. ARIMA process stands for autoregressive (AR), integrated (I), and moving average (MA). Mathematically, ARIMA(p, d, q) can be expressed as

$$\phi L.K^d.y_t = \theta L(\xi_t), \tag{3.8}$$

$$K = 1 - L, \tag{3.9}$$

$$\begin{aligned}
y_t = &(1 + \phi_1)y_{t-1} + (\phi_2 - \phi_1)y_{t-2} + \cdots + \phi_p y_{t-p-1} \\
&+ \xi_t - \theta_1 \xi_{t-1} - \theta_2 \xi_t - 2 - \cdots - \theta_q \xi_{t-q},
\end{aligned} \tag{3.10}$$

where y_t is the time-series variable, d is the difference operator, ξ_t is a white noise with zero mean and infinite variance, and ϕ and θ are coefficients of AR and MA processes. In the statistical package R, there is a function called auto.arima in the forecast package that automatically evaluates such parameters.

Figs. 3.5 and 3.6 illustrate wind speed forecasting for ARMA(1, 1) and ARIMA(2, 1, 1) models, respectively.

Simulations are performed on SPSS 16.0 platform using time-series analyzer. Based on forecast results, we note that ARIMA model performs better than ARMA with error statistics depicted in Table 3.2.

This also reinforces the fact that ARIMA models are superior to ARMA models for forecasting nonstationary time series. Literature survey shows that ARIMA model is extensively used to forecast the wind speed and power and serves as a benchmark model to assess the forecast accuracy of a newly developed or hybrid model. In [14] a variant of ARIMA model, f-ARIMA, is used to predict the day ahead wind forecasts. In this model, f stands for fractional, where the differencing parameter $d \in (-0.5, 0.5)$. Results reveal that f-ARIMA model is 42% more efficient than the persistence model.

3.4 Machine learning-based models

Wind flow depends on atmospheric variables like pressure, humidity, surface roughness, and elevation. The historical behavior plays a pivotal role

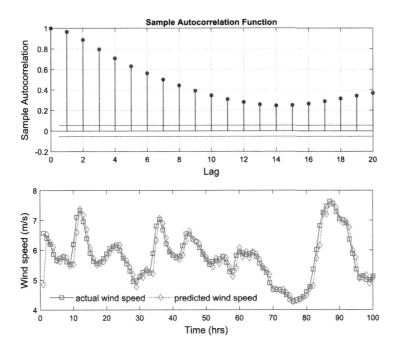

Figure 3.5 Autocorrelation of wind speed and forecasting using ARMA(1, 1) model.

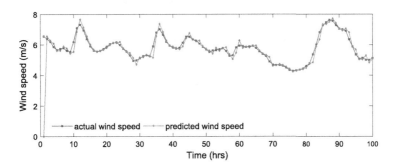

Figure 3.6 Wind speed forecasting using ARIMA(2, 1, 1) model.

Table 3.2 Forecast statistic for ARMA and ARIMA models.

Model	R^2	RMSE	MAE	MAPE
ARMA(1, 1)	0.964	0.222	0.159	3.655
ARIMA(2, 1, 1)	0.970	0.203	0.141	3.229

for short-term predictions. Thus, to solve this issue with nonlinearity, machine learning (ML) models are being used globally to predict the wind speed. Machine learning is a part of broader umbrella called artificial intelligence. The term machine learning focuses on specific task of either classifying, predicting, pattern recognition, clustering, or a combination of some or all of these through a large set of real-time examples.

Machine learning is categorized in two branches, supervised learning and unsupervised learning. Supervised learning is the type of learning where the ML algorithm receives labeled input examples and intends to converge to the best as possible classifier $f: X \rightarrow Y$, so that a new set of examples can be classified, where X is the representative input feature set, and Y is the distinctive labeled class [15]. In case of classification, Y is a set of values either 1 or 0, which corresponds to an individual class. On the other hand, unsupervised learning builds a classifier that essentially does not rely on input examples but learns itself by analyzing the similarities in the input data.

Accurate wind speed forecasting depends on availability of large historical dataset, which is useful in training a supervised ML regression model. Among supervised ML regression methods, support vector machines, artificial neural networks, and extreme learning machine are consistently used.

Fig. 3.7 illustrates everyday examples of machine learning categorized into supervised and unsupervised learning methods.

Classification falls into the category of supervised learning where the individual category of *fruits* as depicted in Fig. 3.7 are treated as labels. On the other hand, in case of unsupervised learning, the clustering problem is addressed via grouping based on similarities found in input data. Unlike supervised learning, in this case, no explicit labels are provided to the algorithm.

The different variants of supervised and unsupervised machine learning are illustrated in Fig. 3.8.

Artificial neural networks are extensively used to forecast wind speed/ power for a duration ranging from 3 min to 6 h. Among ANNs, the most commonly used topology isa multilayered perceptron (MLP). In [8], short-term wind power forecasting for wind farms in Portugal is achieved via a hybrid method of wavelet transform and neural networks (NNs), where initially the wind speed data are broken down into subseries via discrete-wavelet transform, and later it is fed to NN for the training part. Once the training algorithm is through, the data sets are sent for learning stage,

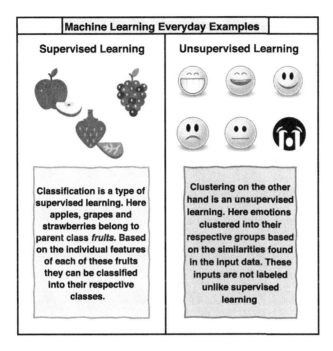

Figure 3.7 Everyday examples of machine learning.

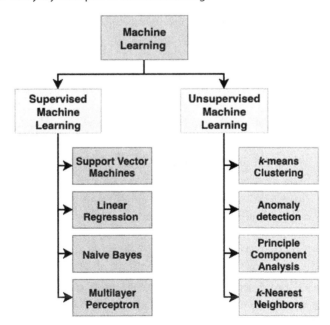

Figure 3.8 Categorization of machine learning algorithms.

where the error minimization takes place between input values and desired values. Usually, back-propagation is used as a learning algorithm [16]. However, back-propagation is a slower technique and so is replaced by the Levenberg–Marquardt algorithm. Results of the neural network wavelet transform (NNWT) approach are compared with ARIMA(1, 2, 1) and NN, where the forecast for 3 h ahead is carried out by taking historical data of the previous 12 h. The MAPE using NNWT approach is found to be 6.97%.

3.5 Hybrid wind forecasting methods

Individual machine learning methods like ANN, SVR, ELM, and GPR are used with signal preprocessing techniques such as the wavelet transform and empirical mode decomposition. Signal decomposition techniques segment the wind speed time series into subseries. These subseries can be used to forecast individually and are aggregated to obtain the final forecast series.

Hybrid models that encompass the features of individual methods have been a core area of research and are illustrated in Fig. 3.9.

Yuan et al. [17] have proposed a hybrid forecasting model employing autoregressive fractionally integrated moving average (ARIFMA) and least-square support vector regression to predict the short-term wind power. Various machine learning-based forecasting models have proven their superiority over time-series models. Among machine-learning methods, the support vector regression estimates a regression function in the spirit of classical support vector machines [18]. Liu et al. [10] have proposed a hybrid short-term wind speed forecasting method using the wavelet transform and support vector machines optimized by the genetic algorithm.

Another hybrid method that studies wind speed forecasting in Mexico uses ARIMA and ANN models is also available [19]. Liu and Tian [20] have used wavelet and wavelet packet along with ANN to predict the wind speed and have compared their findings with standard models like ANFIS, wavelet-RBF, and the persistence method. Xiao et al. [21] have demonstrated combined prediction models employing weight calculation for individual models according to their superiority among various other models. In [22] the authors have used ANN with statistical weight preprocessing technique to predict the coefficient of performance of a ground source heat pump (GCHP) system. Further studies have shown that using optimization methods for selecting model parameters yields better results

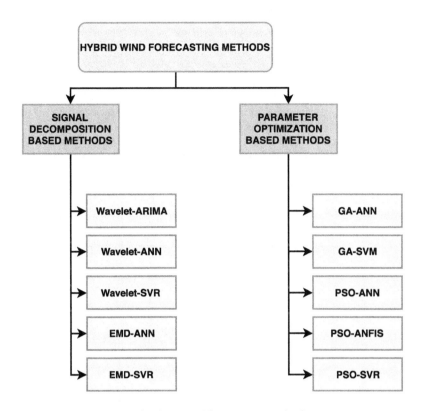

Figure 3.9 Categorization of Hybrid wind forecasting methods.

than individual methods. A commonly used optimization technique is the genetic algorithm owing to its fast convergence. Su et al. [23] have used particle swarm optimization to select optimal model parameters for ARIMA model combined with Kalman filter for wind speed forecasting. Various hybrid techniques have been employed recently to study short-term wind forecasting for wind farms. Signal decomposition methods like wavelet transform and empirical mode decomposition are commonly used for studying the stochasticity in wind speed. Wavelet decomposition along with neural networks is studied by Du et al. and related works, where the prediction accuracy has been enhanced with a multiobjective optimization of hyperparameters [24–28]. The model is then compared with benchmark methods like the persistence method, ARIMA, and generalized neural network. Results reveal that multiobjective optimization–based neural network models outperform the benchmark model and are reliable for handling the short-term wind forecasting needs.

References

[1] M. Alexiadis, P. Dokopoulos, H. Sahsamanoglou, Wind speed and power forecasting based on spatial correlation models, IEEE Transactions on Energy Conversion 14 (3) (1999) 836–842, https://doi.org/10.1109/60.790962.

[2] S.P. Kani, M. Ardehali, Very short-term wind speed prediction: a new artificial neural network–Markov chain model, Energy Conversion and Management 52 (1) (2011) 738–745, https://doi.org/10.1016/j.enconman.2010.07.053.

[3] Y. Jiang, Z. Song, A. Kusiak, Very short-term wind speed forecasting with Bayesian structural break model, Renewable Energy 50 (2013) 637–647, https://doi.org/10.1016/j.renene.2012.07.041.

[4] M. Negnevitsky, P. Johnson, Very short term wind power prediction: a data mining approach, in: 2008 IEEE Power and Energy Society General Meeting – Conversion and Delivery of Electrical Energy in the 21st Century, IEEE, 2008.

[5] Q. Hu, R. Zhang, Y. Zhou, Transfer learning for short-term wind speed prediction with deep neural networks, Renewable Energy 85 (2016) 83–95, https://doi.org/10.1016/j.renene.2015.06.034.

[6] C. Zhang, H. Wei, X. Zhao, T. Liu, K. Zhang, A gaussian process regression based hybrid approach for short-term wind speed prediction, Energy Conversion and Management 126 (2016) 1084–1092.

[7] Z. Liang, J. Liang, L. Zhang, C. Wang, Z. Yun, X. Zhang, Analysis of multi-scale chaotic characteristics of wind power based on Hilbert–Huang transform and Hurst analysis, Applied Energy 159 (2015) 51–61, https://doi.org/10.1016/j.apenergy.2015.08.111.

[8] J. Catalão, H. Pousinho, V. Mendes, Short-term wind power forecasting in Portugal by neural networks and wavelet transform, Renewable Energy 36 (4) (2011) 1245–1251, https://doi.org/10.1016/j.renene.2010.09.016.

[9] A. Conejo, M. Plazas, R. Espinola, A. Molina, Day-ahead electricity price forecasting using the wavelet transform and ARIMA models, IEEE Transactions on Power Systems 20 (2) (2005) 1035–1042, https://doi.org/10.1109/tpwrs.2005.846054.

[10] D. Liu, D. Niu, H. Wang, L. Fan, Short-term wind speed forecasting using wavelet transform and support vector machines optimized by genetic algorithm, Renewable Energy 62 (2014) 592–597.

[11] H. Pousinho, V. Mendes, J. Catalão, A hybrid PSO–ANFIS approach for short-term wind power prediction in Portugal, Energy Conversion and Management 52 (1) (2011) 397–402, https://doi.org/10.1016/j.enconman.2010.07.015.

[12] S. wei Fei, Y. He, Wind speed prediction using the hybrid model of wavelet decomposition and artificial bee colony algorithm-based relevance vector machine, International Journal of Electrical Power & Energy Systems 73 (2015) 625–631, https://doi.org/10.1016/j.ijepes.2015.04.019.

[13] S. Salcedo-Sanz, Á.M. Pérez-Bellido, E.G. Ortiz-García, A. Portilla-Figueras, L. Prieto, D. Paredes, Hybridizing the fifth generation mesoscale model with artificial neural networks for short-term wind speed prediction, Renewable Energy 34 (6) (2009) 1451–1457, https://doi.org/10.1016/j.renene.2008.10.017.

[14] R.G. Kavasseri, K. Seetharaman, Day-ahead wind speed forecasting using f-ARIMA models, Renewable Energy 34 (5) (2009) 1388–1393.

[15] R.F. de Mello, M.A. Ponti, Machine Learning, Springer International Publishing, 2018.

[16] J.L. Paixão, J. Rigodanzo, J.P. Sausen, J.R. Hammarstron, A.R. Abaide, L.N. Canha, M.M. Santos, Wind generation forecasting of short and very short duration using neuro-fuzzy networks: a case study, in: 2017 International Conference on Modern Power Systems (MPS), 2017, pp. 1–6.

[17] X. Yuan, Q. Tan, X. Lei, Y. Yuan, X. Wu, Wind power prediction using hybrid autoregressive fractionally integrated moving average and least square support vector machine, Energy 129 (2017) 122–137.

[18] V.N. Vapnik, The Nature of Statistical Learning Theory, Springer, New York, 2000.

[19] E. Cadenas, W. Rivera, Wind speed forecasting in three different regions of Mexico, using a hybrid ARIMA–ANN model, Renewable Energy 35 (12) (2010) 2732–2738.

[20] H. Liu, H. qi Tian, D. fu Pan, Y. Li fei, Forecasting models for wind speed using wavelet, wavelet packet, time series and artificial neural networks, Applied Energy 107 (2013) 191–208.

[21] L. Xiao, J. Wang, Y. Dong, J. Wu, Combined forecasting models for wind energy forecasting: a case study in China, Renewable & Sustainable Energy Reviews 44 (2015) 271–288.

[22] H. Esen, M. Inalli, A. Sengur, M. Esen, Forecasting of a ground-coupled heat pump performance using neural networks with statistical data weighting pre-processing, International Journal of Thermal Sciences 47 (4) (2008) 431–441.

[23] Z. Su, J. Wang, H. Lu, G. Zhao, A new hybrid model optimized by an intelligent optimization algorithm for wind speed forecasting, Energy Conversion and Management 85 (2014) 443–452.

[24] P. Du, J. Wang, W. Yang, T. Niu, A novel hybrid model for short-term wind power forecasting, Applied Soft Computing 80 (2019) 93–106, https://doi.org/10.1016/j.asoc.2019.03.035.

[25] J. Wang, W. Zhang, Y. Li, J. Wang, Z. Dang, Forecasting wind speed using empirical mode decomposition and Elman neural network, Applied Soft Computing 23 (2014) 452–459.

[26] H. Liu, C. Chen, H. qi Tian, Y. fei Li, A hybrid model for wind speed prediction using empirical mode decomposition and artificial neural networks, Renewable Energy 48 (2012) 545–556.

[27] S. Wang, N. Zhang, L. Wu, Y. Wang, Wind speed forecasting based on the hybrid ensemble empirical mode decomposition and GA-BP neural network method, Renewable Energy 94 (2016) 629–636.

[28] J. Wang, P. Du, T. Niu, W. Yang, A novel hybrid system based on a new proposed algorithm—multi-objective whale optimization algorithm for wind speed forecasting, Applied Energy 208 (2017) 344–360.

CHAPTER 4

Supervised machine learning models based on support vector regression

A vast majority of practically utilized machine methods in fact involve supervised learning techniques. In supervised learning, an algorithm is used to learn an approximate mapping function from the input variable x to the output $y = f(x)$ so that when there are new input data, we can predict the output for that data. Support vector regression which is derived from its parent version is based on an optimization problem. The weights corresponding to each input sample in a training set are obtained. Variants such as Least square support vector regression, Twin support vector regression and ε-Twin support vector regression are presented with a case study for each. In this chapter, we describe such regression models in detail.

4.1 Support vector regression

Support vector regression (SVR) is a type of machine learning regression, which is associated with learning algorithm equipped to analyze historical data for classification and regression. SVR works on the principle of structural risk minimization (SRM) from statistical learning theory [1], [2]. The core idea of this theory is to construct a hypothesis h that yields the lowest true error for the unseen and random sample testing data [3]. Apart from SVR, a universal machine intelligent technique like artificial neural network (ANN) with applications in character recognition, image compression, and stock market prediction is studied [4]. Shirzad et al. [5] have compared the performance of ANN and SVR to predict the pipe burst rate (PBR) in water distribution networks (WDNs) and found ANN to be a better predictor than SVR, but generalization is not consistent with physical behavior. SVR has an advantage over ANN with respect to the number of parameters involved in training phase. The computation time is another important factor for carrying out regression analysis.

Consider a set of training data (historical data) $(x_1, y_1), \ldots, (x_n, y_n) \subset X \times \mathrm{R}$, where X denotes the input feature space of dimension R^n. Let $Y =$

Supervised Machine Learning in Wind Forecasting and Ramp Event Prediction
https://doi.org/10.1016/B978-0-12-821353-7.00015-6
41

$(y_1, y_2, \ldots, y_i)$ denote the set representing the training output or response, where $i = 1, 2, \ldots, n$ and $y_i \in \mathbb{R}$.

4.2 ε-support vector regression

This type of SVR uses an ε-insensitive loss function that intuitively accounts for sparsity similar to SVR by ignoring errors less than ε.

ε-SVR aims to find the linear regressor

$$f(x) = w^T x + b \quad \text{with } w \in X, b \in \mathbb{R}, \tag{4.1}$$

for prediction, where $x \in X$ is the input set containing all the features, w is the weight coefficient related to each input x_i, and b is the bias term.

The objective is to find $f(x)$ with maximum deviation ε from the respective feature sets while being as flat as possible. To achieve the flatness of the desired regressor, the square of the norm of weight vector w needs to be minimized, and the SVR problem is structured in the form of a convex optimization problem [6] as

$$\min \frac{1}{2} \| w \|^2 + C(e^T \chi + e^T \chi^*), \tag{4.2}$$

$$\textbf{subject to} \quad y - w^T x - eb \le e\varepsilon + \chi, \chi \ge 0, \tag{4.3}$$

$$w^T x + eb - y \le e\varepsilon + \chi^*, \chi^* \ge 0,$$

where C is the regularization factor that reflects the trade-off between the flatness of regressor $f(x)$ and the maximum tolerable deviation ε. The value of ε introduces a margin of tolerance where no penalty is imposed on the errors. The larger the ε, the larger the errors. The parameter C controls the amount of influence of the error. The variables χ, χ^* are the slack variables introduced as a soft margin to the tolerable error ε, and e is the vector of ones of appropriate dimension $n \times 1$.

In machine learning regression the problem of overfitting persists, which results in less error in training phase and high error in testing phase. Commonly used regularization techniques include $\mathcal{L}_1$ and $\mathcal{L}_2$ regularization. Mathematically, these are expressed as

$$\mathcal{L}_1: \underset{w}{\arg\min} \text{ loss function} + \lambda \sum_{i=1}^{n} |w_i|, \tag{4.4}$$

$$\mathcal{L}_2: \underset{w}{\arg\min} \text{ loss function} + \lambda \sum_{i=1}^{n} w_i^2. \tag{4.5}$$

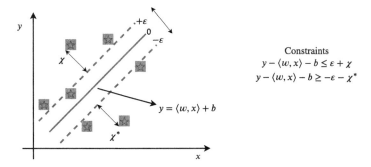

Figure 4.1 Diagrammatic explanation of ε-SVR.

The formulation of SVR is diagrammatically explained in Fig. 4.1, where the "stars" represent the support vectors, green solid line shows the SVR regressor, and blue dashed lines are the hyperplanes with soft limit on the tolerance error ε.

However, this is not always the case, as the feature sets might not be linearly separable. To handle such nonlinearities in the feature sets, the so-called kernel functions are used to transform data to a higher-dimensional space ("kernel trick"). After transformation via a suitable mapping function $\phi : R^n \rightarrow Z$, the data become linearly separable in the target space (high-dimensional space), that is, Z. The inner product $\langle w^T, \phi(x) \rangle$ in the target space can be represented by using a kernel function. Kernel functions are similarity functions that satisfy Mercer's theorem such that $k(x_i, x_j) = \langle \phi(x_i), \phi(x_j) \rangle$ are the elements of the kernel matrix K. Several kernel functions are available in the literature like linear, polynomial of degree d, Gaussian, and the radial basis function (RBF) with bandwidth of the function σ and exponential function.

Fig. 4.2 illustrates the kernel trick used when the input vectors are not linearly separable. This transformation makes the computation of weights and bias vector much easier. Now we look at the dual form of the optimization problem:

$$\min \frac{1}{2}\sum_{i,j=1}^{n}(\alpha_i - \alpha_i^*)^T k(x_i, x_j)(\alpha_j - \alpha_j^*) + e^T \varepsilon \sum_{i=1}^{n}(\alpha_i + \alpha_i^*)$$

$$-\sum_{i=1}^{n} y_i(\alpha_i - \alpha_i^*) \qquad (4.6)$$

$$\textbf{s.t.} \quad e^T \sum_{i=1}^{n}(\alpha_i - \alpha_i^*) = 0, \quad 0 \le \alpha, \quad \alpha^* \le Ce,$$

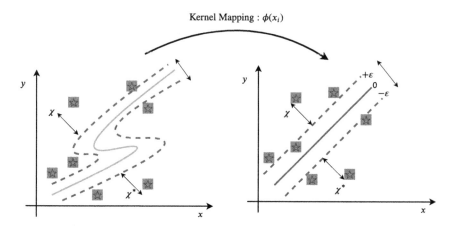

Figure 4.2 Kernel mapping into higher-dimensional space for nonlinear datasets.

where α and α^* represent the positive and negative Lagrange multipliers such that $\alpha_i\alpha_i^* = 0$, $i = 1, 2, \ldots, n$. The regressor $f(x)$ can be written as

$$f(x) = \sum_{i=1}^{n}(\alpha_i - \alpha_i^*)k(x, x_i) + b. \tag{4.7}$$

The complexity of this regressor is independent of the dimensionality of the feature set and only depends on the number of support vectors, which are nothing but the data points that separate the feature sets from each other. However, the performance of the SVR also depends on the choice of a kernel function and helps in reducing the computation time of the regression. A flowchart with a step-by-step implementation of the ε-SVR algorithm is described in Fig. 4.3.

ε-SVR-based regression models are used in several applications as illustrated in Fig. 4.4.

Chen et al. [7] have described a parameter search-based SVR forecasting method for predicting stock prices. The parameter search algorithms like the genetic algorithm, particle swarm optimization, and grid search are used to use the optimal value of SVR hyperparameters. The daily historical trading data of CSI 300 index is chosen for 3 years. The training set comprises of 759 days and remaining 190 days for testing. It is found that GA-SVR-based method yielded minimum RMSE followed by PSO-SVR and grid search. Tao et al. [8] have presented a simulated annealing (SA) based SVR model for forecasting air conditioning load. Hourly data

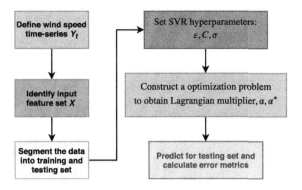

Figure 4.3 Schematic flowchart for ε-SVR algorithm.

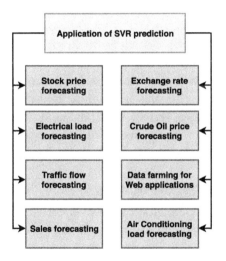

Figure 4.4 Applications of ε-SVR prediction.

from July 2011 to August 2012 are taken. The training set comprises of 10 months from July 2011 to June 2012, whereas the testing set is of 2 months (July–August 2012). A mixed kernel function comprising polynomial kernel and radial basis function is taken. Three models, that is, SA-SVR, VFSA-SVR, and MSA-SVR, are tested, and results reveal that MSA-SVR outperforms VFSA-SVR and SA-SVR. The average time consumed by each of these methods is also computed, and it is found that VFSA-SVR is the fastest of the three models.

In terms of traffic flow forecasting, Hong et al. [9] have carried out a hybrid forecasting model employing ant-colony optimization (ACO) based

Table 4.1 Performance metrics for ε-SVR and persistence model.

Model	RMSE	MSE	R^2	NMSE	U1	U2
ε-SVR	13.63	185.77	1.099	0.0201	0.004	0.2017
Persistence	35.12	1233.41	7.6129	6.8104	0.0749	0.5517

SVR. Continuous ACO is used to optimize σ, C, and ε, and the forecasting results are compared with seasonal ARIMA (S-ARIMA) model. The traffic flow data for the city of Panchiao, Taiwan, are segmented into training and testing data for morning and evening peak hours. In terms of normalized RMSE, CACO-SVR model outperforms S-ARIMA model. Due to better SVR generalization capability and added advantage of parameter optimization, the model holds an upper edge over S-ARIMA. Sales growth forecasting is studied by Wang et al. [10] utilizing improved PSO (IPSO) based SVR method. To assess accuracy of the proposed IPSO-SVR model, it is tested against benchmark models like multiple regression model, back propagation neural network, and direct regression model, and results reveal that IPSO-SVR outperforms all other methods. Overall, hybrid SVR models involving parameter search algorithms are used extensively for real-time forecasting problems. The supervised learning approach significantly improves forecast performance.

Stock price prediction: a case study

Next, we discuss the opening stock price prediction for L&T Infotech keeping the highest price, lowest price, and closing price for a particular day as predictor variables. Daily data for two years are taken from August 21, 2017, to August 20, 2019. A total of 492 samples are segmented into training (400) and testing (92) sets. The hyperparameters are tuned in the range of 2^{-10}–2^{10} for SVR model. The results from Table 4.1 indicate that SVR model outperforms the persistence model in terms of RMSE and R^2 value. Here the problem of overfitting persists, which can be potentially eliminated by using better versions of SVR models discussed in the next section. Fig. 4.5 illustrates the prediction of opening stock prices for both models.

4.3 Least-square support vector regression

Least-square support vector regression (LSSVR) is originally derived from least-square support vector classifiers (LSSVC) proposed in [11], where

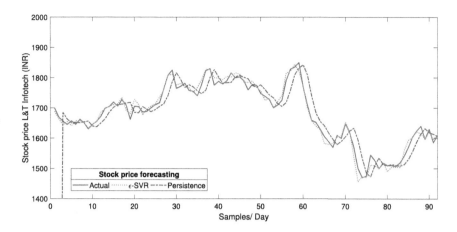

Figure 4.5 Stock price prediction for Larsen & Toubro Infotech.

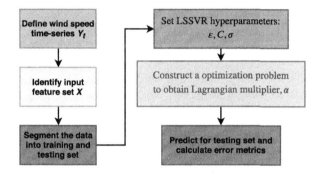

Figure 4.6 Schematic flowchart for LSSVR algorithm.

equality constraints are chosen, and the square of the error term ε is minimized. The LSSVR algorithm is depicted in Fig. 4.6.

The LSSVR regression problem is formulated as

$$f(x) = w^T \phi(x) + b, \tag{4.8}$$

where w is the weight coefficient vector of dimension $n \times 1$, and $x_i \in \mathrm{R}^n$, $y \in \mathrm{R}$. The objective function to be minimized is given as

$$\min \frac{1}{2} \| w \|^2 + \frac{1}{2} \gamma \sum_{i=1}^{n} \varepsilon_i^2 \tag{4.9}$$

$$\textbf{s.t.} \quad y_i = w^T \phi(x_i) + b + \varepsilon_i \quad (i = 1, 2, \ldots, n), \tag{4.10}$$

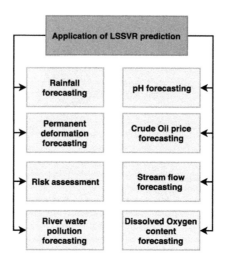

Figure 4.7 Applications of LSSVR prediction.

where γ is the margin parameter, and ε_i is the error term corresponding to each x_i. The optimization problem can be transformed by introducing Lagrange multipliers and is given as

$$L(w, b, \varepsilon, \alpha) = \frac{1}{2} \parallel w \parallel^2 + \frac{1}{2}\gamma \sum_{i=1}^{n} \varepsilon_i^2 - \sum_{i=1}^{n} \alpha_i (w^T \phi(x_i) + b + \varepsilon - y_i). \quad (4.11)$$

The Karush–Kuhn–Tucker (KKT) conditions for the optimization problem (4.11) can be obtained by partially differentiating the Lagrangian function with respect to w, b, ε, α given by

$$\begin{bmatrix} k(x, x^T) + \gamma^{-1} I & e \\ e^T & 0 \end{bmatrix} \begin{bmatrix} \alpha \\ b \end{bmatrix} = \begin{bmatrix} \gamma \\ 0 \end{bmatrix}, \quad (4.12)$$

$$f_{LSSVR}(x) = \sum_{i=1}^{n} \alpha_i k(x, x_i) + b, \quad (4.13)$$

where I is the identity matrix of appropriate dimension. The regressor obtained by LSSVR is given by (4.13) and solves the optimization problem of size smaller than classical ε-SVR, taking less computation time. LSSVR is an improved variant of ε-SVR with a squared penalty term in the loss function and is applied to several branches of science and engineering to obtain accurate forecasts as illustrated in Fig. 4.7.

Liu et al. [12] have studied the parameter optimization methods in tandem with LSSVR for forecasting dissolved oxygen content in transformer oil during incipient faults. The data for dissolved gases are collected from several power companies, and imperialist competition algorithm (ICA) is used to find optimal hyperparameters for LSSVR. The method is compared with several other methods like the back propagation neural network, generalized regression neural network, and radial basis function neural network. The results reveal that ICA-LSSVR-based forecasting yields better results in terms of MAPE and R^2.

In terms of river flow forecasting, Adnan et al. [13] have carried out a hybrid forecasting method based on LSSVR and the gravitational search algorithm (HLSGA). Two catchment areas on the upper Indus basin of Pakistan are selected, and the mean annual data for 32 years are chosen to validate the hybrid method, where n datasets are divided into $n-1$ sets for training. Various input combinations based on the autocorrelation values of the river flow data are used. The results reveal that HLSGA method outperforms model 5 regression tree (M5RT) and multiple linear regression (MLR) in terms of RMSE and MAE. The effect of log transformation on the river flow time series is analyzed, and the forecasting results of log-HLSGA are compared with HLSGA. The log transformation reduces the skewness in the time series and yields better forecasts for all the methods when compared to the original methods.

Power generation prediction: a case study

Next, we discuss the prediction of net hourly power output in megawatts from a combined cycle power plant (CCPP). The principle of CCPP is based on power generation from gas and steam turbines, where variables like temperature (°T), ambient pressure (m Bar), exhaust vacuum (mm of Hg), and relative humidity (%) are taken as predictor variables. The data for this case study can be accessed from UCI machine learning repository [14]. A total of 500 samples are taken and divided into training (400) and testing (100) sets. From Table 4.2 we can observe that LSSVR outperforms the persistence algorithm in terms of RMSE, R^2 (coefficient of determination), and the normalized mean squared error (NMSE). Further, in terms of R^2, the overfitting problem posed by the persistence algorithm is well treated by LSSVR technique.

Fig. 4.8 illustrates the prediction of power generated based on LSSVR and the persistence algorithm.

Table 4.2 Performance metrics for LS-SVR and persistence model.

Model	RMSE	MSE	R^2	NMSE	U1	U2
LSSVR	4.899	24.00	0.888	0.079	0.0054	0.2118
Persistence	20.94	438.48	14.09	14.42	0.0733	0.8720

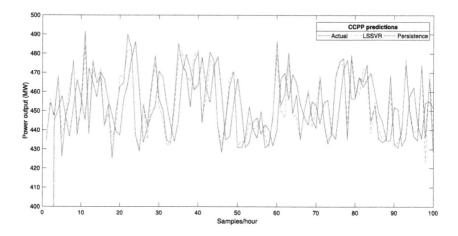

Figure 4.8 Power prediction based on LSSVR and persistence algorithm.

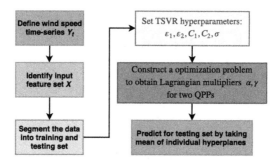

Figure 4.9 Schematic flowchart for TSVR algorithm.

4.4 Twin support vector regression

Peng [15] introduced an efficient solution of the regression through a modified form of support vector machines known as twin support vector regression (TSVR), which derives two nonparallel hyperplanes around the data points. The flowchart depicting the steps involved in the TSVR algorithm is shown in Fig. 4.9.

In a nonparallel plane regressor, two nonparallel functions pertaining to the ε-insensitive down and up-bounds of the unknown regressor are formulated. Similar to ε-SVR, TSVR finds two ε-insensitive functions (up-bound and down-bound regressors) to solve the convex optimization problem having size smaller than the conventional ε-SVR, thus significantly reducing the computation time.

The mathematical formulation of TSVR is

$$\min \frac{1}{2} \sum_{i=1}^{n} (y_i - e\varepsilon_1 - (x_i w_1 + e b_1))^T (y_i - e\varepsilon_1 - (x_i w_1 + e b_1)) \tag{4.14}$$

$$+ C_1 e^T \sum_{i=1}^{n} \xi_i, \qquad \textbf{s.t.} \quad y_i - (x_i w_1 + e b_1) \geq e\varepsilon_1 - \xi_i,$$

$$\min \frac{1}{2} \sum_{i=1}^{n} (y_i - e\varepsilon_2 - (x_i w_1 + e b_2))^T (y_i - e\varepsilon_2 - (x_i w_2 + e b_2)) \tag{4.15}$$

$$+ C_2 e^T \sum_{i=1}^{n} \eta_i, \qquad \textbf{s.t.} \quad (x_i w_2 + e b_2) - y_i \geq e\varepsilon_2 - \eta_i,$$

where $C_1, C_2 > 0$ and $\varepsilon_1, \varepsilon_2 \geq 0$ are the TSVR hyperparameters, and ξ_i, η_i are the slack variables introduced as a soft margin to the error ε in the optimization problem. The dual optimization problem formulation of TSVR is given by introducing the Lagrangian function [15]

$$L(w_1, b_1, \varepsilon_i, \alpha_i, \beta_i) = \frac{1}{2} \sum_{i=1}^{n} (y_i - e\varepsilon_1 - (x_i w_1 + e b_1))^T$$

$$(y_i - e\varepsilon_1 - (x_i w_1 + e b_1)) + C_1 e^T \sum_{i=1}^{n} \xi_i$$

$$- \sum_{i=1}^{n} \alpha_i (y_i - e\varepsilon_1 - (x_i w_1 + e b_1)) - \sum_{i=1}^{n} \beta_i \xi, \tag{4.16}$$

where α_i, β_i $(i = 1, 2, \ldots, n)$ are the Lagrangian multipliers. Let $X = (x_1, x_2, \ldots, x_n)$ denote the set of input vectors, $Y = (y_1, y_2, \ldots, y_n)$ be the set of output vectors, where $y_i \in R$, and α, γ are the Lagrangian multipliers. The KKT conditions for the Lagrangian function utilized in TSVR optimization can be obtained by partially differentiating the function L with respect to $w_1, b_1, \varepsilon, \alpha,$ and β:

$$\begin{cases} \dfrac{\partial L}{\partial w_1} = 0 \Rightarrow -X^T(Y - Xw_1 - eb_2 - e\varepsilon_1) + X^T\alpha = 0, \\[2mm] \dfrac{\partial L}{\partial b_1} = 0 \Rightarrow -e^T(Y - Xw_1 - e\varepsilon_1 - eb_2) + e^T\alpha = 0, \\[2mm] \dfrac{\partial L}{\partial \xi} = 0 \Rightarrow C_1 e^T - \alpha - \beta = 0, \\[2mm] \dfrac{\partial L}{\partial \alpha} = 0 \Rightarrow Y - (Xw_1 + eb_1) \geq e\varepsilon - \xi, \quad \xi \geq 0. \end{cases}$$

Further, the equality constraints for this optimization problem are

$$\alpha^T(Y - (X w_1 + e b_1) - e\varepsilon_1 + \xi) = 0, \quad \alpha = 0,$$
$$\beta^T\xi = 0, \quad \beta \geq 0, \tag{4.17}$$

where $\alpha \in [0, C_1 e]$ for $\beta \geq 0$, which can be written as

$$-\begin{bmatrix} X^T \\ e^T \end{bmatrix}\left((Y - e\varepsilon_1) - \begin{bmatrix} X & e \end{bmatrix}\begin{bmatrix} w_1 \\ b_1 \end{bmatrix}\right) + \begin{bmatrix} X^T \\ e^T \end{bmatrix}\alpha = 0. \tag{4.18}$$

Denoting

$$Q = \begin{bmatrix} X & e \end{bmatrix}, \quad t = Y - e\varepsilon_1, \quad u_1 = \begin{bmatrix} w_1^T & b_1 \end{bmatrix}^T, \tag{4.19}$$

we have

$$-Q^T t + Q^T Q u_1 + Q^T\alpha = 0, \tag{4.20}$$
$$u_1 = (Q^T Q)^{-1} Q^T(t - \alpha).$$

The matrix $Q^T Q$ is always positive semidefinite, meaning that it always has nonnegative eigenvalues. Such matrices may suffer from the problem of "ill-conditioned" matrix, where the condition number might be too large, and computation of its inverse may lead to large numerical errors. To overcome this, a small regularization constant σI is added, where σ is of order 10^{-7} and I is the identity matrix of appropriate dimensions. Combining KKT conditions [15] and optimization problem described by (4.14), the dual problem can be reformulated as

$$\max \; -\frac{1}{2}\alpha^T Q(Q^T Q)^{-1} Q^T\alpha + t^T Q(Q^T Q)^{-1} Q^T\alpha - t^T\alpha \tag{4.21}$$

$$\textbf{s.t.} \; \alpha \in [0, C_1],$$

$$\max \; -\frac{1}{2}\gamma^T Q(Q^T Q)^{-1}Q^T\gamma + m^T Q(Q^T Q)^{-1}Q^T\gamma - m^T\gamma \qquad (4.22)$$

$$\textbf{s.t. } \gamma \in [0, C_2],$$

where $Q = [X\; e]$, $t = Y - e\varepsilon_1$, $m = Y + e\varepsilon_2$, and $u_2 = (Q^T Q)^{-1}Q^T(m - \gamma)$. Eqs. (4.21)–(4.22) refer to the dual of original convex optimization problem, where the size of the former is smaller than classical SVR, thereby making it faster. The final regressor for predicting raw data points is given as

$$f_{TSVR}(x) = \frac{1}{2}((w_1 + w_2)^T x + (b_1 + b_2)). \qquad (4.23)$$

In terms of literature, Gupta et al. [16] presented TSVR model for predicting financial time series where various stock market index prices of companies like Yahoo, Infosys Limited, AT&T, Cisco, Facebook, and Citi Group are undertaken. Performance indices for SVR and TSVR are compared, and results reveal the superiority of TSVR over the classic SVR model. In terms of training time, TSVR outperforms SVR for all datasets. Further, the average rank for SVR and TSVR is also calculated using the Friedman statistic, and it is revealed that TSVR holds an edge over SVR model for nonlinear kernel functions as well. This section describes a twin support vector regression based on twin support vector machines (TSVM). Because of the ability of this method to solve two quadratic programming problems, a final regressor results in considerably lesser computation time.

Wine quality prediction: a case study

The Minho region of Portugal is well known for white and red wines. Based on its several physio-chemical properties, the wine quality can be modeled and compared with benchmark prediction models like the persistence algorithm. The properties like citric acid content (g/dm^3), sulphate content (g/dm^3), density, pH, alcohol content (%), volatile acidity, and fixed acidity are used as predictor variables. Data for red wine are collected from May 2004 to December 2005 and are segmented into training (400) and testing (100) sets. Twin support vector regression and persistence model are used to predict the wine quality, which is a quantitative index ranging from 0 to 10. Fig. 4.10 illustrates a plot between actual and predicted values of wine quality for TSVR and persistence model. Further, Table 4.3 depicts the performance metrics for TSVR and persistence models. TSVR outperforms the persistence model by 7.91% and in terms of R^2 and thus gives

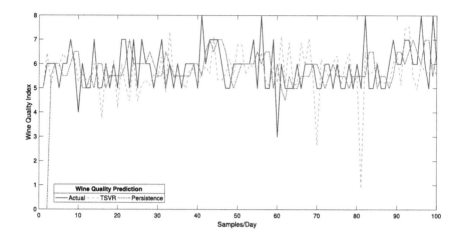

Figure 4.10 Wine quality prediction based on TSVR.

Table 4.3 Performance metrics for TSVR and persistence model.

Model	RMSE	MSE	R^2	NMSE	U1	U2
TSVR	1.0029	1.0058	1.004	1.2351	0.0861	45.82
Persistence	1.0891	1.1861	1.2364	2.1764	0.0921	54.62

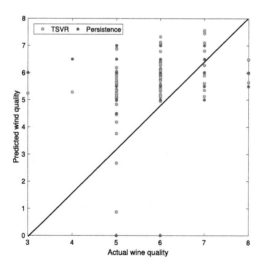

Figure 4.11 Scatter plot for wine quality prediction based on TSVR.

a better value than the persistence model. Fig. 4.11 illustrates a scatter plot for TSVR and persistence model.

4.5 ε-twin support vector regression

Derived from a twin support vector machine, Shao et al. [17] propose a novel regressor, ε-twin support vector regression (ε-TSVR), which determines the pair of ε-insensitive functions by solving two convex optimization problems. In terms of the objective function to be minimized, ε-TSVR considers an added regularization term that solves the ill-conditioning problem of $Q^T Q$ as described in the previous section. The formulation of primal objective functions for ε-TSVR is

$$\min \ \frac{1}{2}C_3(w_1^T w_1 + b_1^2) + \frac{1}{2}\xi^{*T}\xi + C_1 e^T \xi,$$

$$\text{s.t. } Y - (Xw_1 + eb_1) = \xi^*, \tag{4.24}$$

$$Y - (Xw_1 + eb_1) \geq -e\varepsilon_1 - \xi, \xi \geq 0, \tag{4.25}$$

$$\min \ \frac{1}{2}C_4(w_2^T w_2 + b_2^2) + \frac{1}{2}\xi^{*T}\xi + C_2 e^T \eta,$$

$$\text{s.t. } Y - (Xw_2 + eb_2) = \eta^*, \tag{4.26}$$

$$Y - (Xw_2 + eb_2) \geq -e\varepsilon_2 - \eta, \eta \geq 0. \tag{4.27}$$

In this optimization problem, C_1, C_2, ε_1, ε_2 are the hyperparameters that determine the regression performance. The Lagrangian function for these two primal problems can be written as

$$
\begin{aligned}
L(w_1, b_1, \xi, \alpha, \beta) = &\ \frac{1}{2}(Y - (Xw_1 + eb_1))^T (Y - (Xw_1 + eb_1)) \\
&+ \frac{1}{2}C_3(w_1^T w_1 + b_1^2) + C_1 e^T \xi \\
&- \alpha^T (Y - (Xw_1 + eb_1) + e\varepsilon_1 + \xi) - \beta^T \xi,
\end{aligned}
\tag{4.28}
$$

where $\alpha = (\alpha_1, \alpha_2, \ldots, \alpha_n)$ and $\beta = (\beta_1, \beta_2, \ldots, \beta_n)$ are the Lagrangian multipliers. To obtain the dual of the stated primal objective functions, KKT conditions are given by

$$
\begin{cases}
\dfrac{\partial L}{\partial w_1} = 0 \Rightarrow -X^T(Y - Xw_1 - eb_2 - e\varepsilon_1) + X^T \alpha + C_3 w_1 = 0, \\[2mm]
\dfrac{\partial L}{\partial b_1} = 0 \Rightarrow -e^T(Y - Xw_1 - e\varepsilon_1 - eb_2) + e^T \alpha + C_3 b_1 = 0, \\[2mm]
\dfrac{\partial L}{\partial \xi} = 0 \Rightarrow C_1 e^T - \alpha - \beta = 0, \\[2mm]
\dfrac{\partial L}{\partial \alpha} = 0 \Rightarrow Y - (Xw_1 + eb_1) \geq -e\varepsilon - \xi, \quad \xi \geq 0,
\end{cases}
$$

$$\alpha^T(Y - (Xw_1 + eb_1) + e\varepsilon_1 + \xi) = 0. \quad \alpha = 0, \quad \beta^T \xi = 0, \quad \beta \geq 0, \quad (4.29)$$

where $\alpha \in [0, C_1 e]$ for $\beta \geq 0$. These KKT conditions can be written as

$$-\begin{bmatrix} X^T \\ e^T \end{bmatrix} Y + \left(\begin{bmatrix} X^T \\ e^T \end{bmatrix} \begin{bmatrix} X & e \end{bmatrix} + C_3 I \right) \begin{bmatrix} w_1 \\ b_1 \end{bmatrix} + \begin{bmatrix} X^T \\ e^T \end{bmatrix} \alpha = 0, \quad (4.30)$$

$$-Q^T Y + (Q^T Q + C_3 I)u_1 + Q^T \alpha = 0, \quad (4.31)$$

where $Q = \begin{bmatrix} X & e \end{bmatrix}$, $u_1 = \begin{bmatrix} w_1^T & b_1 \end{bmatrix}^T$.

Further, we can write $u_1 = (Q^T Q + C_3 I)^{-1} Q^T (Y - \alpha)$. The dual optimization objective function for the above primal can be written as

$$\max \; -\frac{1}{2}\alpha^T Q(Q^T Q + C_3 I)^{-1} Q^T \alpha^T + Y^T Q(Q^T Q + C_3 I)^{-1} Q^T \alpha$$

$$-(e^T \varepsilon_1 + Y^T)\alpha \quad \textbf{s.t. } \alpha \in [0, C_1]. \; (4.32)$$

Similarly the other dual can be obtained as

$$\max \; -\frac{1}{2}\gamma^T Q(Q^T Q + C_4 I)^{-1} Q^T \gamma^T + Y^T Q(Q^T Q + C_4 I)^{-1} Q^T \gamma$$

$$+(-e^T \varepsilon_2 + Y^T)\gamma \quad \textbf{s.t. } \gamma \in [0, C_2]. \; (4.33)$$

Eqs. (4.32) and (4.33) are the duals of the primal objective optimization function. The feature set X here is linearly separable in the n-dimensional space. The end regressor $f(x)$ is given as

$$f(x) = \frac{1}{2}(f_1(x) + f_2(x)) = \frac{1}{2}((w_1 + w_2)^T x + (b_1 + b_2)), \quad (4.34)$$

as the mean of two functions $f_1(x)$ and $f_2(x)$, which are the up and down regressors across the predicted hyperplane $f(x)$.

Kernel ε-twin support vector regression

Regression via SVR may not be always linear, and thus this study needs to be extended to nonlinear regression. A suitable mapping (also called kernel) function is used to transform the input set into higher dimension using $\phi : R^n \rightarrow R^k$, where k is the dimension of the target space.

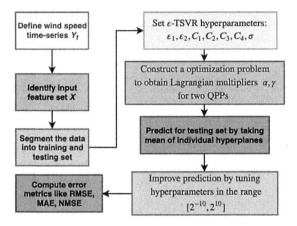

Figure 4.12 Schematic flowchart for ε-TSVR algorithm.

Fig. 4.12 illustrates step by step the procedure of ε-TSVR algorithm for regression.

As in [17], the convex optimization problem considering kernel function $K(X, X^T)$ is given as

$$\min\ \frac{1}{2}C_3(w_1^T w_1 + b_1^2) + \frac{1}{2}\xi^{*T}\xi + C_1 e^T \xi,$$
$$\textbf{s.t.}\ \ Y - (K(X, X^T)w_1 + eb_1) = \xi^*, \tag{4.35}$$
$$Y - (K(X, X^T)w_1 + eb_1) \geq -e\varepsilon_1 - \xi, \xi \geq 0, \tag{4.36}$$
$$\min\ \frac{1}{2}C_4(w_2^T w_2 + b_2^2) + \frac{1}{2}\xi^{*T}\xi + C_2 e^T \eta,$$
$$\textbf{s.t.}\ \ Y - (K(X, X^T)w_2 + eb_2) = \eta^*, \tag{4.37}$$
$$Y - (K(X, X^T)w_2 + eb_2) \geq -e\varepsilon_2 - \eta, \eta \geq 0, \tag{4.38}$$

where C_1, C_2, C_3, C_4 are the hyperparameters for kernel-based ε-TSVR. The duals of the primal optimization problems are given as

$$\max\ -\frac{1}{2}\alpha^T S(S^T S + C_3 I)^{-1}S^T \alpha^T + Y^T S(S^T s + C_3 I)^{-1}S^T \alpha$$
$$-(e^T \varepsilon_1 + Y^T)\alpha \quad \textbf{s.t.}\ \alpha \in [0, C_1], \tag{4.39}$$
$$\max\ -\frac{1}{2}\gamma^T S(S^T S + C_4 I)^{-1}S^T \gamma^T + Y^T S(S^T S + C_4 I)^{-1}S^T \gamma$$
$$+(-e^T \varepsilon_2 + Y^T)\gamma \quad \textbf{s.t.}\ \gamma \in [0, C_2], \tag{4.40}$$

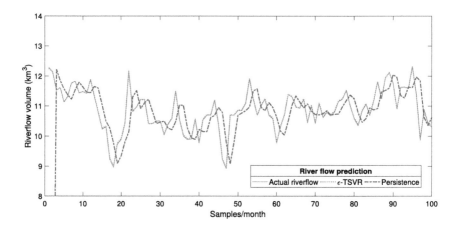

Figure 4.13 Riverflow forecasting based on ε-TSVR.

where $S = [K(X, X^T) \quad e]$, and α, γ are the Lagrangian multipliers, which enable the finding of local maxima and/or minima of the function subject to equality constraints. The end regressor $f_{\varepsilon - TSVR}(x)$ is given as the mean of the two functions:

$$f_{\varepsilon - TSVR}(x) = \frac{1}{2}((w_1^T + w_2^T)K(X, X^T) + (b_1 + b_2)). \qquad (4.41)$$

River flow forecasting: a case study for Nile river

Next, we discuss the river flow volume forecasting based on ε-TSVR model. Monthly mean for river flow volume is collected from https://people.sc.fsu.edu/~jburkardt/datasets/time_series/. A total of 570 samples are taken and are segmented into training (470) and testing (100) sets. The input set for the regression is constructed by wavelet decomposition of river flow time series. The detailed and approximate signals are then treated as predictor variables. Figs. 4.13 and 4.14 illustrate the predicted time-series plots and scatter plots for ε-TSVR and persistence models.

Based on the manual tuning of the hyperparameters, from Table 4.4 we obtain the RMSE 0.0257 of ε-TSVR model and 0.6150 of the persistence model.

In this chapter, we primarily discussed SVR and its variants for regression analysis. These supervised learning models are applied for wind speed time-series datasets from different places around the globe to assess the performance of each model. The forecasting performance in terms of

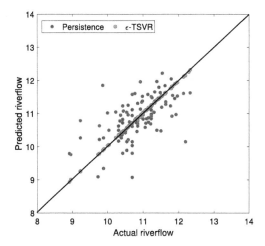

Figure 4.14 Scatter plot for ε-TSVR and persistence algorithm.

Table 4.4 Performance metrics for ε-TSVR and persistence model.

Model	RMSE	MSE	R^2	NMSE	U1	U2
ε-TSVR	0.0257	0.0006	0.9653	0.013	0.0012	0.3440
Persistence	0.6150	0.3782	5.5652	6.771	0.0845	6.6146

overfitting and underfitting is further discussed in subsequent chapters. Further, the superiority of hybrid forecasting models has led to an increased research in the field of wind speed forecasting.

References

[1] V. Vapnik, A. Lerner, Pattern recognition using generalized portrait method, Automation and Remote Control 24 (1963) 774–780.
[2] C. Cortes, V. Vapnik, Machine Learning 20 (3) (1995) 273–297.
[3] T. Joachims, Text categorization with support vector machines: learning with many relevant features, in: Machine Learning: ECML-98, Springer Berlin Heidelberg, 1998, pp. 137–142.
[4] B. Schölkopf, A. Smola, K.-R. Müller, Kernel Principal Component Analysis, Lecture Notes in Computer Science, Springer Berlin Heidelberg, 1997, pp. 583–588.
[5] A. Shirzad, M. Tabesh, R. Farmani, A comparison between performance of support vector regression and artificial neural network in prediction of pipe burst rate in water distribution networks, KSCE Journal of Civil Engineering 18 (4) (2014) 941–948.
[6] V.N. Vapnik, The Nature of Statistical Learning Theory, Springer New York, 2000.
[7] J. Chen, H. Chen, Y. Huo, W. Gao, Application of SVR models in stock index forecast based on different parameter search methods, Open Journal of Statistics 07 (02) (2017) 194–202.

[8] Y. Tao, H. Yan, H. Gao, Y. Sun, G. Li, Application of SVR optimized by Modified Simulated Annealing (MSA-SVR) air conditioning load prediction model, Journal of Industrial Information Integration 15 (2019) 247–251.

[9] W.-C. Hong, Y. Dong, F. Zheng, C.-Y. Lai, Forecasting urban traffic flow by SVR with continuous ACO, Applied Mathematical Modelling 35 (3) (2011) 1282–1291.

[10] X. Wang, J. Wen, S. Alam, X. Gao, Z. Jiang, J. Zeng, Sales growth rate forecasting using improved PSO and SVM, Mathematical Problems in Engineering 2014 (2014) 1–13.

[11] J. Suykens, J. Vandewalle, Neural Processing Letters 9 (3) (1999) 293–300.

[12] J. Liu, H. Zheng, Y. Zhang, X. Li, J. Fang, Y. Liu, C. Liao, Y. Li, J. Zhao, Dissolved gases forecasting based on wavelet least squares support vector regression and imperialist competition algorithm for assessing incipient faults of transformer polymer insulation, Polymers 11 (1) (2019) 85.

[13] R.M. Adnan, X. Yuan, O. Kisi, R. Anam, Improving accuracy of river flow forecasting using LSSVR with gravitational search algorithm, Advances in Meteorology 2017 (2017) 1–23.

[14] Uci machine learning repository: combined cycle power plant data set, https://archive.ics.uci.edu/ml/datasets/Combined+Cycle+Power+Plant, 2019. (Accessed 22 August 2019).

[15] X. Peng, TSVR: an efficient twin support vector machine for regression, Neural Networks 23 (3) (2010) 365–372.

[16] D. Gupta, M. Pratama, Z. Ma, J. Li, M. Prasad, Financial time series forecasting using twin support vector regression, PLoS ONE 14 (3) (2019) e0211402.

[17] Y.-H. Shao, C.-H. Zhang, Z.-M. Yang, L. Jing, N.-Y. Deng, An ε-twin support vector machine for regression, Neural Computing & Applications 23 (1) (2012) 175–185.

CHAPTER 5

Decision tree ensemble-based regression models

Historical data in machine learning are a boon for data analysts round the world. In terms of regression models, benchmark models like the persistence algorithm and ARIMA described in Chapter 3 fail to express the nonlinear trend in wind speed time series. Machine learning models with kernel mapping capability are widely applicable in such cases.

In terms of accuracy, machine learning models based on supervised learning give good generalization performance, which is the ability of a machine learning model to adapt to an unseen data. However, the need for ensemble-based forecasts arrives from the problem of overfitting, which occurs in models like support vector regression, multiple linear regression, and neural networks. We will discuss the overfitting scenarios in subsequent chapters considering real-time datasets. Decision trees aim to arrive at a decision of two types: continuous (regression) and discrete (classification).

5.1 Random forest regression

Proposed by Brieman [1], a random forest is an ensemble method that generates something akin to a forest of trees from a given training sample. Ensemble-based models are far more accurate than a single method owing to advantages like capturing linearity and nonlinearity of time series obtained from individual methods.

A random forest begins with splitting the input features into a group of subsets that essentially form a tree. A particular tree is characterized by a node that leads to a number of branches, as depicted in Fig. 5.1.

Similar to hyperparameters (σ and C) in SVR, in random forest regression, the number of trees and the number of random features in each tree decomposition are the parameters that decide the performance of regression. At each decision tree, a fitting function is created, which acts on the random features selected. Finally, at the end of the training process a random forest model is created. It is worth noting that during the training process, each tree is created from randomly selected input vectors and thus is called a "random" forest. Mathematically, the estimated output of a

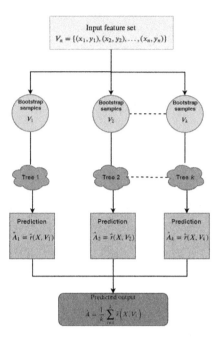

Figure 5.1 Block diagram for random forest regression model.

random forest regression is given as

$$\hat{A} = \frac{1}{k}\sum_{i=1}^{k}\hat{r}\left(X, V_i\right),\qquad(5.1)$$

where $\hat{r}\left(X, V\right)$ is the representative tree at the end of training process, X is the set of input feature vectors, T is the collective set representing an input–output pair $V_i = (x_1, y_1), (x_2, y_2), \ldots, (x_n, y_n)$. The predicted output is averaged over k decision trees. An additional advantage of random forest regression is its insensitivity to noise due to uncorrelated trees via differential sampling of inputs. Fig. 5.1 illustrates the block diagram of random forest regression model. Decision trees are sensitive to the data on which they are trained. Changing the training data can change the predictions.

Meanwhile a common problem of overfitting persists in machine learning regression models when a well trained model captures the noise component as well. To reduce the complexities posed by overfitting, a random forest makes a compromise between flexible and inflexible models. In the training phase, each regression tree repeatedly draws a sample from the feature set. This ensures that even though the tree may possess a high variance, the overall variance of the forest may be low. A random forest works on the principle of bagging, which combines the predictions from different tree models to give an overall insight to the data under training. This also helps to reduce the potential overfitting caused by supervised machine learning models. In case of bagging the principle of negative correlation learning (NCL) is also applicable to regression-based problems [2]. In NCL a penalty term is added while computing the error of an individual sample in an ensemble. The modified error function for an individual sample i is given as

$$E_i = \frac{1}{N} \sum_{i=1}^{N} \left(F_i(X) - y_i \right)^2 + \frac{1}{N} \sum_{i=1}^{N} \beta p_i, \qquad (5.2)$$

where $F_i(X)$ is the predicted output for ith sample, y_i is the actual value, $\beta \in [0,1]$ is the term for control of the strength of the penalty, and p_i is the correlation function.

Random forest regression is applied in several branches of engineering. Since random forests work well with missing data, variables like forest fires, rain flow, and soil erosion can be accurately predicted. Oliveira et al. [3] have used random forest (RF) regression to predict forest fires in the Mediterranean region of Europe. The data for fire density is collected from European Fire Database for the months from June to September. Variables like average elevation, minimum temperature, maximum temperature, and surface roughness are chosen as predictors. The predictions from RF model are compared with linear regression. The results reveal that RF model outperforms linear regression indicating a nonlinear relationship among the predictors and forest fires. Gong et al. have discussed a comparison between RF model and linear regression for predicting international roughness index (IRI). The data for roughness index and predictors are average precipitation, average freeze index, initial IRI measurement, and annual average daily traffic. Data of 11000 samples are taken and divided into training (80%) and testing (20%) set. The hyperparameters of RF model are optimized using a grid search method. The results in terms of MSE and R^2 reveal that RF model performs better than the linear regression model.

Smith et al. [4] discussed multiple linear regression (MLR) and random forest regression to predict concentration levels of various chemicals. The nonlinearity among multiple variables is accounted by the random forest model, and the results reveal that although MLR performed better than RF model, there are some cases where the predictive performance of RF model is found much better. Yuchi et al. [5] explored a random forest model and multiple linear regression for predicting the concentration of a particulate matter. The particulate matter $PM_{2.5}$ is predicted based on several predictor variables like outdoor $PM_{2.5}$ concentration, season, the number of gears, and the number of cleaners deployed. A total of 447 samples are assessed and a cross validation technique with 10 folds is used for all the models. Individual models like MLR and RF result in R^2 values of 0.502 and 0.498, respectively, whereas a blended model accounting MLR and RF gave an R^2 value of 0.815, indicating its superiority in predicting the indoor $PM_{2.5}$ concentration.

Zahedi et al. [6] presented an RF model for predicting the solid particle erosion in elbow pipe fittings. Predictor variables like pipe diameter, particle size, liquid viscosity, and material hardness are chosen for this purpose. The performance of RF model is compared with benchmark model like linear regression, support vector machines, and neural network. The results indicate a good agreement in predicted and experimental values with RF model. It is also observed that the RF model overpredicts the erosion rate, which is possibly due to influence of different predictor variables over each other. We now discuss two case studies for random forest model that deal with rainfall and crude oil price prediction.

5.1.1 Monthly rainfall prediction: a case study

To test the effectiveness of the random forest model against benchmark models like persistence algorithm, a case study for monthly rainfall prediction in the state of Himachal Pradesh, India, is carried out. Monthly rainfall data from 1901–2017 are taken, and a total of 1404 samples are segmented into training and testing sets. As far as inputs to the model are concerned, the rainfall time series is decomposed into various low- and high-frequency components using wavelet transform. The predictors for this random forest model include detail components and the fifth approximate signal, where the latter represents low-frequency component. The regression is carried out with 800 training samples and remaining 604 for testing.

Table 5.1 depicts the performance metrics for a random forest model. In terms of RMSE, a value of 25.14 indicates a high error when compared to the persistence algorithm and ARIMA model. This poor performance is depicted by a low SSR/SST value, indicating that for the current dataset, the quality of predictors in terms of wavelet decomposition is not sufficient to carry out regression analysis as per RF model.

Table 5.1 Performance metrics for random forest model.

RMSE	MSE	SSR/SST	SSE/SST	IOA	U1	U2
25.1418	632.0196	0.7130	0.0800	0.999	0.0984	0.5905

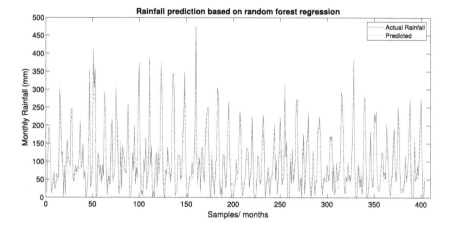

Figure 5.2 Rainfall prediction for Himachal Pradesh as per random forest method.

Fig. 5.2 illustrates the rainfall prediction based on the random forest model for Himachal Pradesh, India.

5.1.2 Crude oil price prediction: a case study

Crude oil is an important trade commodity whose price is subjected to market volatility. Predictions made ahead in time help the traders and oil companies across the globe to plan their imports and exports for economic reasons. The crude oil price prediction is carried out based on a random forest model and the persistence algorithm. Daily prices are modeled as a time series acquired for a duration of approximately 8 years (July 15, 2011, to August 15, 2019) and illustrated in Fig. 5.3. Autocorrelation and partial autocorrelation functions indicate a quantitative relationship between lagged elements of a time series. A total of 2106 samples are segmented into training (1500) and testing (606) sets. The predictor variables for the RF model include detail signals and approximate signals obtained from wavelet transform of price time series. The simulations for RF model are carried out in R software with 1000 trees. See Fig. 5.4.

Table 5.2 depicts the performance metrics for crude oil price prediction. In terms of RMSE, RF model outperforms the persistence algorithm by 52.51%. It is also important to note that SSR/SST value for the persistence

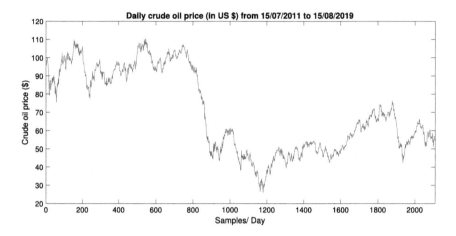

Figure 5.3 Time-series plot for crude oil price.

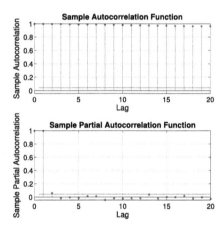

Figure 5.4 Autocorrelation and partial autocorrelation function for crude oil price.

Table 5.2 Performance metrics for random forest model for crude oil prediction.

| Model | RMSE | MSE | SSR/SST |
	SSE/SST	U1	U2
RF	0.5486	0.3009	0.9200
	0.0045	0.0046	0.4925
Persistence	1.1553	1.3347	1.1625
	0.1340	0.0254	1.1235

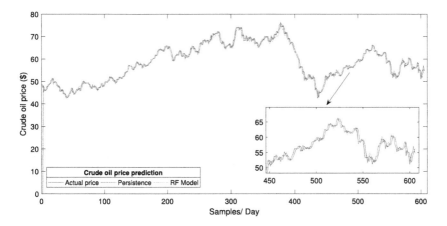

Figure 5.5 Crude oil price prediction based on random forest regression.

algorithm is greater than 1, which creates a problem of overfitting. The problem of overfitting data points is a common phenomenon in machine learning methods as well. In terms of Theil's U1 and U2 statistics, a lower value in RF model indicates the model superiority over the benchmark persistence model. Fig. 5.5 illustrates the prediction plots for RF model and the persistence algorithm.

5.2 Gradient boosted machines

Gradient boosted machines (GBMs) is an ensemble-based regression method, that takes into account the loss of the previously fit decision tree. In a GBM model, various weak learners are combined to arrive at an accurate model. Boosting-based regression trees are highly powerful to predict accurate response values owing to the advantages posed by an appropriate boosting algorithm. A particular boosting algorithm strengthens the tree model by minimizing the inaccuracies of weak models in the form of loss function [7]. Thus for a boosting algorithm, the main task is reducing the error posed by weak learners in each iteration. The final predicted response is the linear combination of fit trees. An advantage of GBM over conventional regression models is that it avoids overfitting. In case of GBM the learning rate of each decision tree determines the predictive performance. A higher learning rate indicates less number of trees required for prediction and vice versa [8]. Based on a cross-validation experiment, an ideal number of trees required along with learning rate can be obtained. Since GBM is a collective model that combines all the individual weak models, the aim for an individual model is reducing the error generated by a loss function. A commonly used loss function is an $\mathcal{L}_2$ function that minimizes the sum of squared errors between the predicted and actual values. Consider k such weak models that account for prediction of a variable y given a feature set $x = [x_1, x_2, \ldots, x_n]$, which mathematically can be expressed as

$$\hat{y} = \sum_{i=1}^{k} f_i(x),\tag{5.3}$$

where $f_i(x)$ represents each weak learner that is collectively trained to improve the prediction results. The loss function is given as

$$\mathcal{L}_2 = \frac{1}{N} L(y_i, \hat{y}_i), \quad L(y_i, \hat{y}_i) = \sum_{i=1}^{N} (y_i - \hat{y}_i)^2,\tag{5.4}$$

where $L(y_i, \hat{y}_i)$ is a loss function based on squared errors for N observations. The aim of a GBM regression technique is to minimize the $\mathcal{L}_2$ loss function. However, the $\mathcal{L}_2$ loss function is more sensitive to outliers and can reduce the robustness of the model. To ensure a high robustness, the Huber loss function is used, which can be expressed as

$$L(y, \hat{y}) = \begin{cases} \frac{1}{2}(y - \hat{y})^2, & |y - \hat{y}| \le \delta, \\ \delta(|y - \hat{y}| - \delta/2), & |y - \hat{y}| > \delta, \end{cases}\tag{5.5}$$

where δ represents the slope. The concept of gradient boosted machine revolves around the gradient descent and boosting algorithm. For optimizing the hyperparameters in GBM, a gradient descent algorithm is used, which minimizes the cost function taking into account the negative of the gradient. Let us consider an $\mathcal{L}_2$ loss function with its gradient of with respect to predicted sample $\hat{y}_i$ given as

$$\frac{\partial L(y, \hat{y})}{\partial \hat{y}_i} = \frac{\partial}{\partial \hat{y}_i} \sum_{i=1}^{N} (y_i - \hat{y}_i)^2 = 2(y_i - \hat{y}_i)\frac{\partial}{\partial \hat{y}_i}(y_i - \hat{y}_i) = -2(y_i - \hat{y}_i).\tag{5.6}$$

The gradient here reflects that while tracking the minimum point of the loss function, the GBM actually tracks the residual vector $y - \hat{y}$. Similarly, for $\mathcal{L}_1$, the MAE can be tracked by finding the gradient of loss function

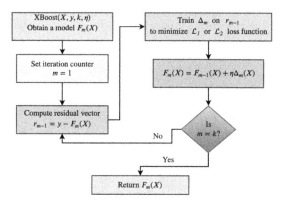

Figure 5.6 XBoost algorithm for gradient boosted machines.

given as

$$\frac{\partial L(y, \hat{y})}{\partial \hat{y}_i} = \frac{\partial}{\partial \hat{y}_i} \sum_{i=1}^{N} |y_i - \hat{y}_i| = -sign(y_i - \hat{y}_i). \quad (5.7)$$

The accuracy of a GBM model depends on the number of weak learners and the learning rate. Further, the essence of a GBM model lies in the boosting technique, which combines all the weak models in a stagewise manner and then in a parallel manner, as seen in the random forest model. Mathematically, the GBM model in recursive form is given as

$$F_m(x) = F_{m-1}(x) + \eta \Delta_m(x), \quad (5.8)$$

where η represents the learning rate, and Δ_m refers to a regression model fitted to the residuals. Fig. 5.6 illustrates the *XBoost* algorithm used for obtaining the best regressor.

In our wind speed prediction simulations, the source codes for random forest and GBM are presented in Appendices A.4 and A.5, respectively.

Song et al. [9] presented a GBM model for predicting clathrate hydrate phase equilibria in salts. The inputs to the model include 11 variables in terms of gases like CH_4, CO_2, and N_2 and salts KCl and NaCl. Two models for deep neural network (DNN) and gradient boosted regression trees with pressure and temperature as outputs are used. Results reveal that the GBM model outperforms DNN for both temperature and pressure as inputs to the model. Further, the GBM model based on temperature as output gave better performance than the GBM model based on pressure.

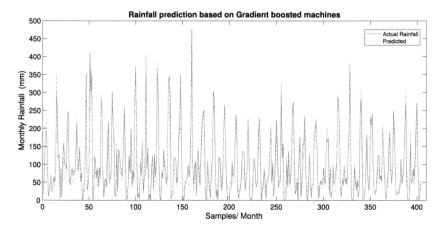

Figure 5.7 Rainfall prediction for Himachal Pradesh, India, based on gradient boosted machines.

Miskell et al. [10] discussed a GBM model for predicting the short-term peak concentration of $PM_{2.5}$ particles for Christchurch, New Zealand. Variables like temperature, relative humidity, wind speed, wind gust, and air pressure are used as predictors in forecast model.

On similar lines a binary classification of high peaks of $PM_{2.5}$ is also carried out using four different GBM models with different interval times. In the field of solar power forecasting, GBM is studied by Persson et al. [11], where multisite solar power forecasting is carried out. Historical data of solar power in Japan from April 14, 2014, to February 28, 2015, are taken and segmented into training (75%) and testing (25%) sets. Scenarios like short-term and long-term prediction are carried out keeping in mind the importance of numerical weather prediction. It is observed that for long-term power predictions the relative importance of numerical weather prediction increases. The normalized RMSE values vary between 0.1 and 0.137 for GBM model when compared to benchmark models like the persistence algorithm, autoregressive model, and climatology.

5.2.1 Monthly rainfall prediction: a case study

Similar to random forest regression, the rainfall prediction is carried out using gradient boosted machines. The predictor set, that is, detail and approximate signals for the rainfall time series, remains the same. From the time–series plot shown in Fig. 5.7, the variability in rainfall is high and represents a nonlinear nature. Model simulations are run in R by segmenting

Table 5.3 Performance metrics for gradient boosted machines model.

RMSE	MSE	SSR/SST	SSE/SST	U1	U2
25.0592	627.55	0.8084	0.0795	0.0975	0.4546

Table 5.4 Performance metrics for GBM model for crude oil prediction.

Model	RMSE	MSE	SSR/SST
	SSE/SST	U1	U2
GBM	0.6486	0.4207	0.8420
	0.0085	0.0076	0.6925
Persistence	1.1553	1.3347	1.1625
	0.1340	0.0254	1.1235

the dataset into training (800) and testing (606) sets. An optimal number of trees is determined through cross validation.

Table 5.3 depicts the performance metrics for rainfall prediction. Results reveal that GBM outperforms RF model in terms of RMSE and SSR/SST. Both RF and GBM model give similar predictive performance and can be used for time series with highly nonlinear nature.

5.2.2 Crude oil price prediction: a case study

Prices of crude oil change everyday and are often studied from prediction point of view by various oil companies. Common prediction models may fail to deliver accurate values for oil prices under which a company may suffer losses in a long run. Similar to RF model, the predictions are performed for GBM model with predictors as detail and approximate signals of the price time series. The time series is divided into training and testing sets in a similar fashion like that in the random forest model case.

In terms of RMSE as depicted by Table 5.4, GBM is an inferior model compared to RF model and gives a low SSR/SST value. However, the prediction performance is found to be much better than that of the persistence algorithm, where the problem of overfitting prevails as indicated by a high SSR/SST value.

Fig. 5.8 illustrates the prediction time series for crude oil price and is tested against the persistence model.

This chapter highlights decision tree ensemble-based regression models for time-series prediction. Random forest and gradient boosted machines use the principle of bagging and boosting, respectively, to arrive at the best prediction. The use of decision tree models for regression is best suited

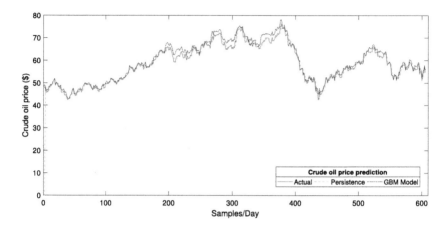

Figure 5.8 Crude oil price prediction based on gradient boosted machines.

when the historical data are large in size. The principle advantage of decision tree models is their ability to combine several weak models to give a strong and robust model capable of giving a better prediction performance when compared to a single model. In the subsequent chapters, we discuss the decision tree ensemble models for ramp event prediction for different onshore and offshore wind farm sites.

References

[1] L. Breiman, J.H. Friedman, R.A. Olshen, C.J. Stone, Classification and Regression Trees, Routledge, 2017.
[2] Y. Liu, X. Yao, Ensemble learning via negative correlation, Neural Networks 12 (10) (1999) 1399–1404.
[3] S. Oliveira, F. Oehler, J. San-Miguel-Ayanz, A. Camia, J.M. Pereira, Modeling spatial patterns of fire occurrence in Mediterranean Europe using multiple regression and random forest, Forest Ecology and Management 275 (2012) 117–129.
[4] P.F. Smith, S. Ganesh, P. Liu, A comparison of random forest regression and multiple linear regression for prediction in neuroscience, Journal of Neuroscience Methods 220 (1) (2013) 85–91.
[5] W. Yuchi, E. Gombojav, B. Boldbaatar, J. Galsuren, S. Enkhmaa, B. Beejin, G. Naidan, C. Ochir, B. Legtseg, T. Byambaa, P. Barn, S.B. Henderson, C.R. Janes, B.P. Lanphear, L.C. McCandless, T.K. Takaro, S.A. Venners, G.M. Webster, R.W. Allen, Evaluation of random forest regression and multiple linear regression for predicting indoor fine particulate matter concentrations in a highly polluted city, Environmental Pollution 245 (2019) 746–753.
[6] P. Zahedi, S. Parvandeh, A. Asgharpour, B.S. McLaury, S.A. Shirazi, B.A. McKinney, Random forest regression prediction of solid particle erosion in elbows, Powder Technology 338 (2018) 983–992.

[7] D. Solomatine, D. Shrestha, AdaBoost.RT: a boosting algorithm for regression problems, in: 2004 IEEE International Joint Conference on Neural Networks (IEEE Cat. No.04CH37541), IEEE, 2004.

[8] J. Elith, J.R. Leathwick, T. Hastie, A working guide to boosted regression trees, Journal of Animal Ecology 77 (4) (2008) 802–813, https://doi.org/10.1111/j.1365-2656.2008.01390.x.

[9] Y. Song, H. Zhou, P. Wang, M. Yang, Prediction of clathrate hydrate phase equilibria using gradient boosted regression trees and deep neural networks, Journal of Chemical Thermodynamics 135 (2019) 86–96.

[10] G. Miskell, W. Pattinson, L. Weissert, D. Williams, Forecasting short-term peak concentrations from a network of air quality instruments measuring PM2.5 using boosted gradient machine models, Journal of Environmental Management 242 (2019) 56–64.

[11] C. Persson, P. Bacher, T. Shiga, H. Madsen, Multi-site solar power forecasting using gradient boosted regression trees, Solar Energy 150 (2017) 423–436.

CHAPTER 6

Hybrid machine intelligent wind speed forecasting models

Modern day grid reliability and security are highly dependent on accurate wind speed and power forecasts. Whereas the nonlinear nature of wind speed poses challenges in forecasting via traditional methods, machine learning-based hybrid models adequately address this issue. In this chapter, we use a hybrid model based on wavelet decomposition technique and several variants of SVR. To determine the best model, performance metrics are calculated for real-time wind speed datasets. ε–twin support vector regression (ε-TSVR) and twin support vector regression (TSVR) are used to forecast short-term wind speed in tandem with classical SVR and LSSVR, and its performance is compared with the benchmark persistence model. We also evaluate the performance of these models for a larger dataset based on wind sites in USA and India.

6.1 Introduction

Research in the field of wind forecasting has tremendously increased for the hybrid models based on machine learning [1]. Candenas and Rivera [2] demonstrated ARIMA-ANN based forecasting model, where for a fixed prediction horizon, wind forecasting is carried out. Liu et al. [3] presented a support vector machine and genetic algorithm (GA) based hybrid forecasting method using the wavelet decomposition transform for fragmenting the wind speed time-series to eliminate any potential stochastic variation. Zhang et al. [4] presented a hybrid technique based on Gaussian process regression (GPR) and autoregression (AR) and compared their results with SVM, ANN, and the persistence algorithm. Mi et al. [5] explored a hybrid method that encompasses the wavelet decomposition transform, extreme learning machine, and outlier correction technique to forecast the multistep wind speed. The wavelet and wavelet packet decomposition eliminates noisy component from the wind series, and the extreme learning machine provides multistep forecast on the subsignals obtained from the decomposition technique.

Supervised Machine Learning in Wind Forecasting and Ramp Event Prediction
https://doi.org/10.1016/B978-0-12-821353-7.00017-X

Li et al. [6] presented a combined method based on constant weight and variable weight for short-term wind speed prediction. Jiang et al. [7] proposed a novel method for short-term wind prediction based on modeling the fluctuations caused by adjacent wind turbines, and the selected inputs are given to a v–SVM model. Azimi et al. [8] presented a feature selection model based on k-means cluster and a multilayer perceptron neural network for predicting short-term wind speed. Jiang et al. presented a correlation-based discrete wavelet transform (DWT), least-square support vector machine (LSSVM), and generalized autoregressive conditional heteroscedastic (GARCH) method for short-term wind speed prediction. Correlation coefficients among different subseries are used to assess the inputs to the LSSVR models for wind speed prediction [9].

Liu et al. [10] discussed a modified Broyden–Fletcher–Goldfarb–Shanno (BFGS) neural network and wavelet transform-based signal processing technique for short-term wind speed prediction and validated the same for four wind speed datasets. Correlation coefficients are determined for each subseries obtained after wavelet decomposition for assessing their relative importance. Tian et al. [11] presented a multiobjective forecasting algorithm wherein data preprocessing technique is based on a complementary ensemble empirical mode decomposition (EEMD), variational mode decomposition, and sample entropy. The proposed model is further validated for eight wind speed datasets, and results reveal the superiority of the model when compared to benchmark models. However, the only limitation of this model is the time consumed in intermediate stages. Three variables, wind speed, electrical load, and electricity price, are predicted using an Elman neural network (ENN), whose weights and hyperparameters are optimized by a modern dragonfly algorithm as presented by Wang et al. [12]. Since the time series for the three variables is highly nonlinear, signal processing techniques such as the wavelet transform, empirical mode decomposition, and ensemble empirical mode decomposition are widely applied in the allied areas of forecasting.

Du et al. [13] presented a multistep ahead prediction based on the whale optimization algorithm (WOA)-LSSVR technique and have applied the same to forecast the wind speed, electricity price, and electrical load. The proposed technique is validated for six datasets from Singapore, China, and Australia. Existing benchmark regressors like the generalized regression neural network (GRNN) and back propagation neural network (BPNN) are used for a comparative analysis. In terms of performance metrics like root mean squared error and mean absolute error, the proposed WOA-LSSVR model outperforms GRNN and BPNN.

Wang et al. [12] presented a hybrid wavelet neural network (WNN) model-based multiobjective sine–cosine algorithm (MOSCA) optimization technique. The optimization problem is based on multiobjective sine–cosine functions. The candidate solutions are first initialized with some value and are allowed to converge or diverge in a given search space. A modified complementary ensemble empirical mode decomposition (MCEEMD) is implemented to solve the issues posed by simple preprocessing techniques such as the ensemble empirical mode decomposition (EEMD). The model proposed by Wang et al. is also evaluated for assessing the robustness and stability for wind speed prediction as a highly nonlinear time series can cause difficulties in accurate prediction. The model based on MOSCA-WNN is applied to predict each subseries, and the results are aggregated. The proposed model is compared with WNN, GRNN, ARIMA, and the persistence model.

Complexity in the form of time consumption can be a serious issue with prediction models, and the same can be addressed with variants of SVR model. In a power system, events like wind ramps and power transients, which can be represented in the form of time series, can be analyzed in a coherent manner by decomposing the signal into the time-frequency domain as indicated by wavelet transform [14]. Throughout this chapter, we discuss the wavelet transform decomposition carried out based on the db4 filter, which is helpful in a smooth and localized decomposition.

A hybrid model based on the wavelet transform and empirical mode decomposition and variants of SVR. A benchmark model like the persistence algorithm is used to compare the performance metrics of the hybrid prediction model. Further, the regularization is carried out on the hybrid models to observe their effect on SSR/SST value. This chapter is segmented as follows: Section 6.2 describes application of wavelet transform to short-term wind speed prediction. Further, Section 6.3 highlights the problem formulation for short-term wind speed forecasting. In Section 6.4, results are discussed for the wavelet transform-based hybrid model, followed by an EMD-based hybrid model for different wind farm sites in Section 6.5.

6.2 Wavelet transform

Signal processing techniques have been used in tandem with machine learning methods to improve the forecast accuracy and eliminate the stochastic variations in the time-series. Signal transforms like the Fourier transform, wavelet transform, and wavelet packet decomposition are the common algorithms used. A major drawback of using the Fourier transform in wind speed decomposition is the loss of information with respect to time scale, which is overcome by the wavelet transform. The wavelet transform captures information of a signal in both time and frequency scales. Since temporal variations hold greater importance in wind speed time-series analysis, the Fourier transform is not preferred.

The idea of wavelet transform, first put forward by a geophysicist Jean Moret in 1982 for seismic wave analysis, can be categorized as the continuous wavelet transform (CWT) and discrete wavelet transform (DWT). The continuous wavelet involves the continuous scaling and time shifting of the mother wavelet. The high scaling (low-pass filter) gives approximate information about the signal, whereas low scaling (high-pass filter) gives a detailed information of the signal. Computationally, DWT is richer than CWT due to which the former finds more use in signal processing. Mathematically, CWT and DWT are given as

$$B(a, b) = \frac{1}{\sqrt{a}} \int_{-\infty}^{+\infty} r(x)\phi\left(\frac{x-b}{a}\right), \tag{6.1}$$

$$B(u, v) = 2^{-u/2} \sum_{t=0}^{N-1} r(t)\phi\left(\frac{t - v \cdot 2^u}{2^u}\right), \tag{6.2}$$

where $r(t)$ is the wind speed time series, N is its length, ϕ is the mother wavelet function, and the scaling and translation parameters are functions of integers u and v. First, the original wind speed time series is decomposed into low- and high-frequency components. The primary objective of the wavelet transform is to collect the vital information from the wind speed time series. With respect to wind forecasting, the appropriate decomposition signals are selected as inputs to the forecasting model. The wavelet filter can be chosen from a group of wavelet families such as Haar, Daubechies, Mexican hat, and Coiflet [15]. The WT process involves a successive decomposition of approximation signal at each stage, as shown in Fig. 6.1.

The two signals obtained at each decomposition stage are approximate signals (A5, containing low-frequency components) and detail signals (D1, ..., D5 with high frequencies), which together form a matrix of input features. The wind speed is the output used in short-term wind forecasting algorithm (SVR and its variants), as depicted in Fig. 6.2.

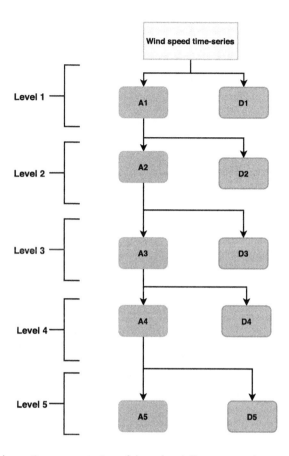

Figure 6.1 Schematic representation of three-level discrete wavelet transform.

6.3 Framework of hybrid forecasting

The present study involves short-term wind speed prediction utilizing a hybrid wavelet transform and an SVR-based method. Hybrid methods are more efficient compared to individual methods as far as the ability to filter stochastic volatility is concerned. The error in wind speed prediction depends on the prediction duration or horizon. For market clearing operations and economic load dispatch, usually short-term wind speed prediction ranging from 30 minutes to 3 hours is a preferred choice. Modern day electricity markets have moved on from regulated markets to deregulated markets, which allow several participants to enter in the bidding process to buy retail electricity from the generation companies. The wind forecasting is carried out using the hybrid model, that is, wavelet-SVR,

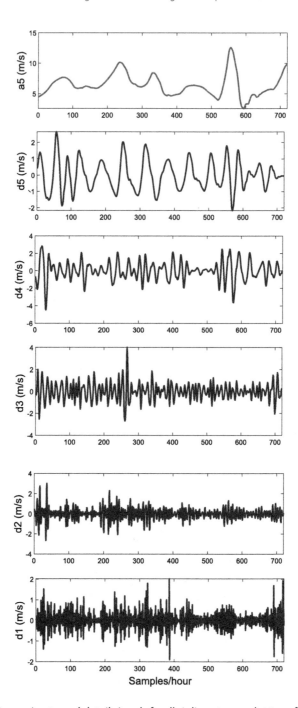

Figure 6.2 Approximate and detail signals for db4 discrete wavelet transform.

wavelet-LSSVR, wavelet-TSVR, and wavelet-ε-TSVR. The forecasting accuracy is evaluated by computing various performance metrics like root mean squared error (RMSE), mean absolute error (MAE), sum of squared residuals (SSR), sum of squared deviation of testing samples (SST), sum of squared error of testing samples (SSE), index of agreement (IOA), and Theil's U1 and U2 statistics [16]. Mathematically, these metrics are expressed as

$$\text{RMSE} = \left[\frac{1}{n} \sum_{i=1}^{n} (\hat{x}_i - x_i)^2 \right]^{1/2}, \quad \text{MAE} = \left[\frac{1}{n} \sum_{i=1}^{n} |\hat{x}_i - x_i| \right], \qquad (6.3)$$

$$\text{SSR/SST} = \frac{\sum_{i=1}^{n}(\hat{x}_i - \bar{x})^2}{\sum_{i=1}^{n}(x_i - \bar{x}_i)^2}, \qquad \text{SSE/SST} = \frac{\sum_{i=1}^{n}(\hat{x}_i - x_i)^2}{\sum_{i=1}^{n}(x_i - \bar{x})^2}, \qquad (6.4)$$

$$\text{IOA} = 1 - \sum_{i=1}^{n} (\hat{x}_i - x_i)^2 \Big/ \sum_{i=1}^{n} (|\hat{x}_i - \bar{x}| + |x_i + \bar{x}|)^2,$$

$$\text{U1} = \sqrt{\frac{1}{n} \times \sum_{i=1}^{n} (\hat{x}_i - x_i)^2} \Big/ \left(\sqrt{\frac{1}{n} \times \sum_{i=1}^{n} x_i^2} + \sqrt{\frac{1}{n} \times \sum_{i=1}^{n} \hat{x}_i^2} \right),$$

$$\text{U2} = \sqrt{\frac{1}{n} \times \sum_{i=1}^{n} ((x_{i+1} - \hat{x}_{i+1})/x_i)^2} \Big/ \sqrt{\frac{1}{n} \times \sum_{i=1}^{n} ((x_{i+1} - \hat{x}_i)/x_i)^2},$$

where $\hat{x}_i$, x_i, and $\bar{x}$ are the predicted, actual, and mean values of the testing samples. Theil's U1 and U2 statistics are important metrics, often used in finance and banking sector to determine the forecast accuracy and quality. The statistic U1 determines how accurate the forecast is, whereas U2 determines the quality of the forecast. Fig. 6.3 shows the block diagram of forecasting through the hybrid wavelet-SVR method.

Description of datasets

To test the hybrid wavelet-SVR, wind farm sites from Spain, Western Massachusetts (USA), South Dakota (USA), Victoria (Australia), and India are chosen with their descriptive statistics listed in Table 6.1.

Fig. 6.4 shows the wind speed variations for these wind farm sites:

- Paxton, MA: The wind site is located in western Massachusetts with 42°18′11.6″ and 71°53′50.9″ as its coordinates. The wind speed is measured every 10 minutes with cup anemometers installed at a height of 78 m above the ground. The wind speed data ranges from January 1, 2011, to January 7, 2011, 22:30 h.

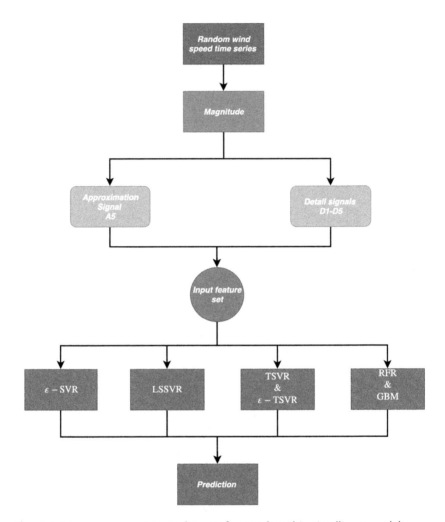

Figure 6.3 Forecasting using wavelet transform and machine intelligent models.

Table 6.1 Descriptive statistics for wind speed at various wind farm sites.

Wind farm (Dataset)	Max (m/sec)	Min (m/sec)	Mean (m/sec)	Std Dev
Sotavento, Spain (A)	13.23	0.41	4.6072	1.9395
Paxton, MA (B)	14.39	0.35	6.9209	2.3734
Blandford, MA (C)	13.73	0.30	6.0553	2.1242
Bishop & Clerks, MA (D)	13.31	0.36	6.7065	2.5923
Beresford, SD (E)	15.06	0.58	5.4729	2.9828
AGL Macarthur (F)	9.05	1.92	6.2926	1.5035
Muppandal (G)	8.48	0.71	4.8878	1.4641

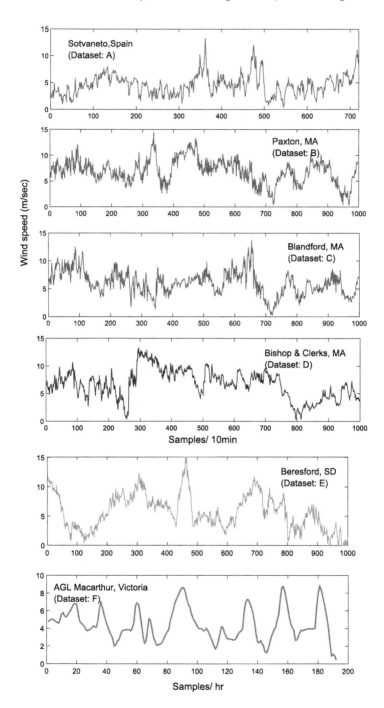

Figure 6.4 Wind speed for datasets A through F.

- Sotavento, Spain: The wind farm is located in Sotavento, Galicia, Spain, with latitude 43°21′35.9″ and longitude −7°52′47.9″. The dataset chosen is for the month of October 2017, where the wind speed is measured hourly.
- Blandford, MA: Blandford is situated at 42.223° N and 72.968° E with wind speed recorded at a height of 60 m above the ground with a cup anemometer at every 10 minutes. The wind speed data ranges from January 1, 2011, to January 7, 2011, 22:30 h.
- Bishop & Clerks, MA: Wind monitoring site is located at 41.574° N and 70.249° E with anemometer installed at height of 15 m above ground. The data ranges from January 1, 2011, to January 7, 2011, 22:30 h and is recorded every 10 minutes.
- Beresford, South Dakota: The wind site is located at 43.088° N and 96.786° E and ranges from March 1, 2006, 22:20 h to March 8, 2006, 20:50 h. Wind speed is recorded every 10 minutes at 20 m height.
- AGL Macarthur, Victoria, Australia: This wind farm is located at 38.049° S and 142.190° E with 420 MW installed capacity featuring 140 V112-3.0 Vestas wind turbines. Hourly wind speed data are taken from February 26, 2019, 00:00 h to March 5, 2019, 23:00 h [17].
- Muppandal, Kanyakumari, India: Located in Kanyakumari, Tamil Nadu; it has a capacity of 1500 MW. Wind speed data for the month of January 2019 are recorded at 10 minute intervals [18].

The autocorrelation function (ACF) plots depict the correlation of time-series samples with itself at different lag instants. Mathematically, ACF at lag instant k for a time series S_t is expressed as

$$r_k = \frac{\sum_{t=1}^{N-k} \left(S_t - \overline{S} \right) \left(S_{t+k} - \overline{S} \right)}{\sum_{i=1}^{N} \left(S_t - \overline{S} \right)^2}. \tag{6.5}$$

Fig. 6.5 illustrates the periodicity of the wind speed time series for all datasets.

For datasets A to E, we find that lag instants of 1 and 2 are significantly dominant, indicating strong correlation. However, the autocorrelation for dataset F is negative for lag order 7.

6.4 Results and discussion

A hybrid model is built on wavelet decomposition technique and a machine intelligent SVR model, where 80% of data are used for training, and

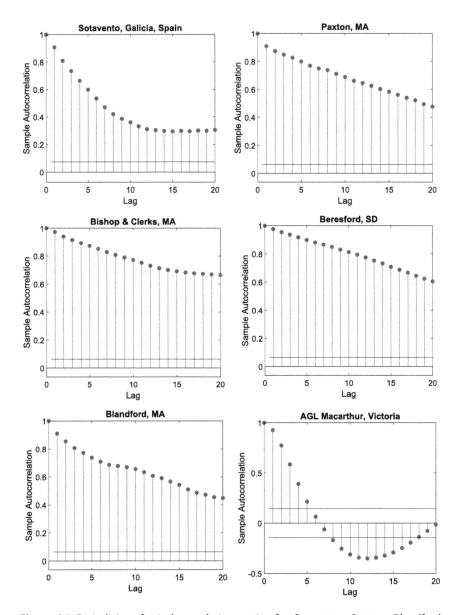

Figure 6.5 Periodicity of wind speed time series for Sotavento, Paxton, Blandford, Bishop & Clerks, Beresford, and AGL Macarthur.

the rest for testing. TSVR and ε-TSVR forecasting models are evaluated vis-a-vis ε-SVR and LS-SVR models. For ε-TSVR, we assume that the regularization factors $C_1 = C_2$ and $C_3 = C_4$. Similarly for TSVR, we select $C_1 = C_2$. Since the regularization parameter sets a soft limit on the tolerance error, selecting an optimal value of C has significant impact on forecast accuracy. Further, since overfitting during training phase causes low accuracy, regularization becomes an important aspect of machine learning regression.

The kernel function used for building the regression models is the radial basis function (RBF) with bandwidth σ: $k(x, x_i) = \exp\left(-\frac{\|x - x_i\|^2}{2\sigma^2}\right)$. The hyperparameters C_1, C_2, C_3, and C_4 along with RBF bandwidth σ are chosen from a set 2^i, $i = -9, -8, \ldots, 9, 10$. Optimal parameters can be tuned manually or by grid search algorithm. Datasets related to four wind farm sites labeled A, B, C, and D are chosen to test the performance of hybrid forecasting model. Dataset A consists of 720 samples out of which 80% (576) are used for training process and 20% (144) are used for testing. Similarly, for datasets B, C, and D, 800 samples are used for training and 200 for testing.

Tables 6.2 and 6.3 depict various performance indices for wavelet-based hybrid SVR models. For dataset A, ε-TSVR and TSVR outperformed ε-SVR by 41.95% and 84.25%, respectively, in terms of RMSE. Similarly, for datasets B, C, and D, ε-TSVR outperforms ε-SVR by 3.537%, 63.03%, and 59.60%, respectively, in terms of RMSE. Among all the models, ε-TSVR and TSVR outperform LS-SVR and ε-SVR quantitatively in terms of RMSE and MAE for all the datasets. In ε-SVR the size of the Lagrangian matrix is large and uses inequality constraints; however, LS-SVR uses equality constraints and is faster in computation.

Further, TSVR is a mixture of ε-SVR and LS-SVR and uses two quadratic programming formulations with better generalizing capability. LS-SVR spends the minimum processor time due to a smaller-sized optimization problem. ε-TSVR takes less time than classical ε-SVR and TSVR, and among the four datasets, B, C, and D take more or less the same computation time for the respective models. The ratios SSR/SST and SSE/SST give an estimate of goodness of fit. The SSR/SST ratio value greater than 1 implies overfitting during training process, which is not desirable during the testing phase. Among all regressors, TSVR obtains optimal SSR/SST and SSE/SST ratios. An optimal SSR/SST ratio is close to 1 but not greater than 1. When overfitting occurs, SSR/SST takes a value greater than 1, which ultimately reduces forecast accuracy. Forecasting is assessed from statistical point of view by determining the index of agreement (IOA) and Theil's U1 and U2 statistics for all the models.

Table 6.2 Performance metrics of wavelet-ε-SVR, LS-SVR, TSVR, and ε-TSVR.

Dataset	Model	RMSE	MAE	SSR/SST	SSE/SST
		IOA	U1	U2	CPU time
A	ε-SVR	0.1423	12.254	0.9434	0.064
		1.0000	0.0233	1.1281	5.7964
	LS-SVR	0.1097	8.4668	0.9417	0.0038
		1.0000	0.0028	0.9665	0.5924
	TSVR	0.0224	1.4877	0.9821	0.0008
		1.0000	0.0011	0.9496	3.1864
	ε-TSVR	0.0823	7.6699	0.9694	0.0021
		1.0000	0.00116	0.0150	2.4421
	Persistence	0.7241	71.41	1.0192	0.2925
		0.9991	0.0913	1.1331	0.3024
B	ε-SVR	0.0424	6.8172	1.0055	0.0003
		1.0000	0.0138	0.0645	11.244
	LS-SVR	0.0329	4.6550	0.9862	0.0001
		1.0000	0.00340	0.0466	0.8165
	TSVR	0.0148	2.0444	0.9965	0.00003
		1.0000	0.0008	0.0072	6.2343
	ε-TSVR	0.0409	6.8764	0.9884	0.0003
		1.0000	0.0021	0.0130	3.6897
	Persistence	0.7825	66.31	1.0197	0.1437
		0.9998	0.0731	1.0000	0.2560

From Table 6.2 we observe that ε-TSVR and TSVR models outperform ε-SVR, LS-SVR, and persistence models in terms of Theil's U1 and U2 statistics, thereby indicating that the accuracy of the two models is superior to the rest. Fig. 6.6 shows the forecasting results of the four variants of SVR for four wind farm sites. The forecast accuracy of ε-TSVR and TSVR is further tested using the Diebold–Mariano (DM) test. The DM statistic test assumes a null hypothesis wherein two forecasting models have similar accuracy [19].

Consider an actual time series y_t and forecasted time series $\hat{y}_{1t}$ and $\hat{y}_{2t}$. The error between actual and forecasted time series is $e_{it} = \hat{y}_{it} - y_t$. Mathematically, the DM test can be expressed as

$$H_0: e_{1t} - e_{2t} = 0, \quad H_1: e_{1t} - e_{2t} \neq 0, \quad \forall t; \quad (6.6)$$

H_0 and H_1 represent the null and alternative hypotheses.

Table 6.3 Performance metrics of wavelet- ε-SVR, LS-SVR, TSVR, and ε-TSVR.

Dataset	Model	RMSE IOA	MAE U1	SSR/SST U2	SSE/SST CPU time
C	ε-SVR	0.1791	24.211	0.9777	0.0151
		0.9980	0.0453	1.3804	12.3536
	LS-SVR	0.0733	8.9438	0.9929	0.0027
		1.0000	0.0021	0.0231	0.8962
	TSVR	0.0100	1.2674	0.9976	0.0004
		1.0000	0.0003	0.0046	6.2235
	ε-TSVR	0.0662	8.7964	1.0127	0.0022
		1.0000	0.0040	0.0635	3.7219
	Persistence	0.6939	54.95	1.0501	0.2934
		0.9998	0.0734	1.0769	0.0818
D	ε-SVR	0.1901	28.8003	1.0538	0.0174
		1.0000	0.0230	0.8751	11.041
	LS-SVR	0.1427	20.5274	0.9798	0.0098
		1.0000	0.0129	0.6855	1.2148
	TSVR	0.0740	9.4829	0.9816	0.0026
		1.0000	0.0050	0.0746	6.7960
	ε-TSVR	0.0768	12.1187	0.9507	0.0028
		1.0000	0.0047	0.0254	9.0181
	Persistence	0.4366	34.71	1.0535	0.1156
		0.9998	0.0587	0.6667	0.0925

Null hypothesis states that the errors for two different forecast series are the same, and the alternative hypothesis contradicts it. In the present case, we compare the DM statistic of TSVR (Test 1) and ε-TSVR (Test 2) against the classical ε-SVR model. The test is carried out at 1% significance level for datasets A, B, C, and D, and the results are highlighted in Table 6.4.

Figs. 6.7 and 6.8 show the variation of SSR/SST ratio with RBF bandwidth σ and regularization factor C for different SVR variants.

Thus by rejecting the null hypothesis from the DM test we observe that both TSVR and ε-TSVR models have significant forecast superiority over ε-SVR model, proving the robustness of the hybrid SVR model and its variants over the persistence model.

The ratio SSR/SST estimates whether the training data have been over trained or not. ε-TSVR and TSVR show better variation of SSR/SST ratio for testing samples than classical ε-SVR and LS-SVR with σ (keeping C

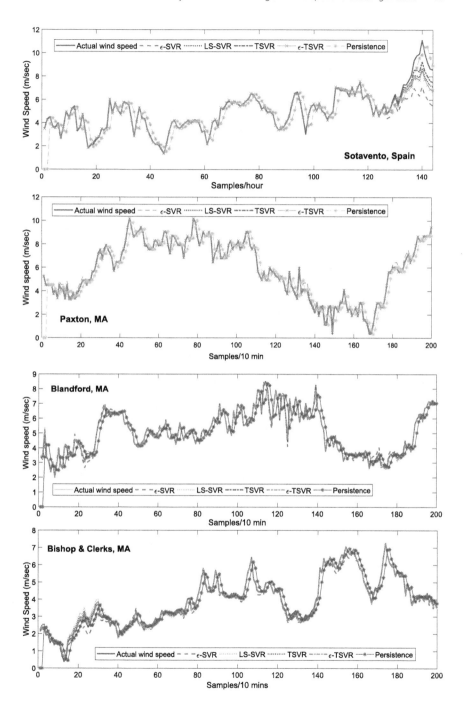

Figure 6.6 Forecasts for ε-SVR, LSSVR, TSVR, ε-TSVR, and persistence model.

Table 6.4 Diebold–Mariano test for datasets.

Dataset	Diebold–Mariano Statistic	
	Test 1	Test 2
A	10.7291	9.7084
B	7.6321	7.4852
C	5.2699	5.2398
D	6.9036	6.6344

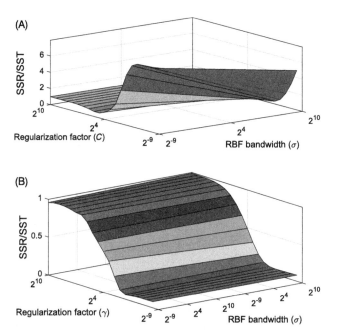

Figure 6.7 Variation of SSR/SST with RBF bandwidth σ and regularization factor (C) for (A) ε-SVR, (B) LS-SVR.

constant) and C (keeping σ constant). As we increase σ (from 2^{-9} to 2^{10}), the value of SSR/SST increases from 0 to 1 for ε-TSVR and TSVR and remains constant for LS-SVR. However, for ε-SVR, the SSR/SST value first decreases and then increases after $\sigma = 2^4$.

To further validate the effect of a larger dataset on our hybrid model, we select the wind speed data from Blandford, MA (Dataset:C), and Muppandal, Kanyakumari (Dataset:G). The training set comprises of 4000 and 3000 samples for datasets C and G, respectively, and the testing set consists

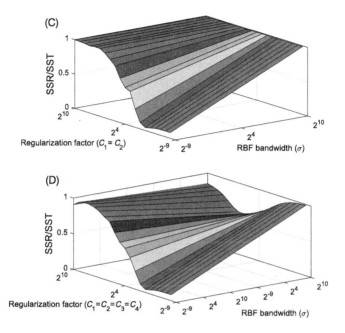

Figure 6.8 Variation of SSR/SST with RBF bandwidth σ and regularization factor (C) for (C) TSVR and (D) ε-TSVR.

of 1000 samples. The forecasting performance is depicted in Table 6.5 and is illustrated in Fig. 6.9.

From Table 6.5 we observe that ε-TSVR and TSVR perform significantly better than ε-SVR and LS-SVR in terms of RMSE and MAE. In terms of the computation speed, ε-TSVR saves 93% and 81% of time compared to ε-SVR for datasets C and G, respectively.

6.5 Empirical mode decomposition-based SVR variants for wind speed prediction

In this section, we discuss a hybrid model based on empirical mode decomposition (EMD) and variants of support vector regression (SVR). Just like wavelet transform, EMD is a signal preprocessing technique for assessing the volatility of a time series. Fig. 6.10 illustrates a flowchart for ARIMA-EMD-based hybrid wind forecasting model. The wind speed prediction is carried out in two parts. First, the linear characteristics are predicted based on an ARIMA model, and second, the nonlinear nature is handled by machine learning models like support vector regression and its variants.

Table 6.5 Performance metrics for a larger dataset.

Dataset	Model	RMSE	MAE	SSR/SST	SSE/SST
	ε-SVR	0.0416	33.7801	1.0079	0.0001
	LS-SVR	0.0194	12.7811	1.0013	0.00003
	TSVR	0.0036	2.2739	0.99994	0.00001
	ε-TSVR	0.0127	8.9721	0.9985	0.00001
	Persistence	0.8553	640.65	0.9826	0.0755
C	**Model**	**IOA**	**U1**	**U2**	**CPU time**
	ε-SVR	1.0000	0.0031	0.1440	911.4211
	LS-SVR	1.0000	0.0014	0.0489	13.3221
	TSVR	1.0000	0.00002	0.0084	355.011
	ε-TSVR	1.0000	0.00094	0.1371	61.0762
	Persistence	1.0000	0.0633	0.7813	0.3024
Dataset	**Model**	**RMSE**	**MAE**	**SSR/SST**	**SSE/SST**
	ε-SVR	0.0283	20.3760	0.9823	0.0008
	LS-SVR	0.0170	12.6992	0.9851	0.00003
	TSVR	0.0143	9.6658	0.9871	0.00002
	ε-TSVR	0.0157	10.9721	0.9910	0.00001
	Persistence	0.2053	73.4500	1.0213	0.0523
G	**Model**	**IOA**	**U1**	**U2**	**CPU time**
	ε-SVR	1.0000	0.0041	0.6416	347.304
	LS-SVR	1.0000	0.0026	0.5163	5.9329
	TSVR	1.0000	0.0022	0.00485	70.2648
	ε-TSVR	1.0000	0.0165	0.1241	63.3087
	Persistence	1.0000	0.0318	0.6053	0.0131

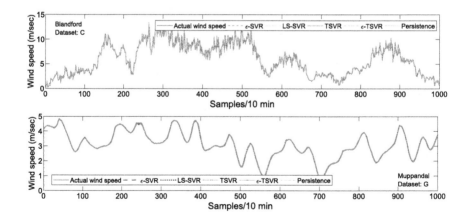

Figure 6.9 Forecasting results for larger datasets.

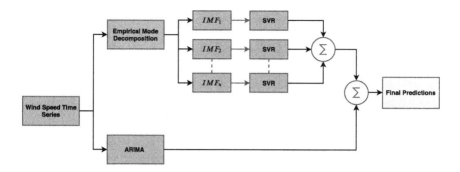

Figure 6.10 Flowchart for ARIMA-EMD-based hybrid wind forecasting model.

The steps for determining the intrinsic mode function (IMFs) and residue of a time series are the following:

- For a given time-series $a(t)$, locate local maximum and minimum and interpolate to form $a_u(t)$ and $a_l(t)$ as upper and lower envelopes.
- Work out the mean of upper and lower envelopes $a_m(t) = \frac{a_u(t) + a_l(t)}{2}$.
- Obtain a detailed component $a_d(t)$ by subtracting the mean component from the original time series as $a_d(t) = a(t) - a_m(t)$.
- Obtain first IMF and residue if $a_m(t)$ and $a_d(t)$ fulfill anyone of the stopping criteria. Stopping criteria: (i) mean envelope developed by maxima and minima should be zero, and (b) number of zero crossings and extrema should differ by one or zero.
- Repeat the previous steps until you obtain the first IMF and residue.

Fan et al. [20] discussed an EMD-SVR-based hybrid model for predicting electrical load in Australian markets where particle swarm optimization is used to determine optimal SVR parameters. Results reveal that the model proposed by Fan et al. performs better when compared to benchmark models like PSO-BP and PSO-SVR. In the field of predicting of ship motion and wave height, Duan et al. [21,22] studied an EMD-based autoregressive (AR)-SVR model, where the nonlinearity in ship motion and wave height is captured by an EMD process. The results reveal significant improvement achieved by EMD-AR-SVR model over AR and SVR models. Xiang et al. [23] explored an SVR-ANN model based on empirical mode decomposition for predicting rainfall in China. Three different datasets are taken, and SVR is tested for short-term predictions, whereas ANN is tested for long-term predictions. The results reveal significant improvement in terms of RMSE and MAE for EMD-SVR and EMD-ANN methods when compared to traditional methods. Yu et al. [24] presented a model based on

EMD-SVR for prediction of frequency spectrum, and the results are compared with autoregressive model and conventional SVR model.

A frequency spectrum data of 300 samples is segmented into training (250) and testing (50) sets. RMSE values of 0.1310, 0.1148, and 0.0673 for AR, SVR, and EMD-SVR models, respectively, indicate the superiority of EMD-based models for predicting the nonstationary time series like that of frequency spectrum. In terms of stock price forecasting, Yang et al. [25] discussed an integrated model based on ARIMA–EMD–SVR for four markets NASDAQ, TSEC, SP500, and NIKKEI 225. Data from January 1, 2010, to December 31, 2010, are used, and 90% of the data are utilized for training the SVR model. Results reveal that ARIMA–EMD-based SVR models give better prediction performance when compared to AR, SVR, and EMD–SVR models. Further, the robustness of the models is also evaluated in terms of a metric called the direction symmetry.

We now discuss the ARIMA–EMD-based SVR models for short-term wind speed prediction. The hybrid model illustrated in Fig. 6.10 involves linear prediction based on ARIMA model and nonlinear prediction based on EMD–SVR model. The lag orders of ARIMA model are determined based on ACF function of the wind speed time series. We consider four datasets Bishop & Clerks, Blandford, and Paxton located in western Massachusetts and Anholt, offshore wind farm site in Denmark, and the wind speed data for these sites can be accessed from the URL (http://www.soda-pro.com). The wind speed time series is measured at every 10 minutes and 1000 samples, out of which 800 are investigated for training SVR model, and the remaining 200 for testing. To begin with, a wind speed time-series is decomposed into five intrinsic mode functions (IMFs) including one residue. These IMFs are then predicted individually, and their predictions are then aggregated. The hyperparameters of SVR model and its variants are tuned manually in the range 2^{-10}–2^{10}. Figs. 6.11 and 6.12 illustrate the forecasting results for four different datasets.

The ARIMA–EMD-based SVR variants are assessed based on the error metrics such as RMSE, mean squared error (MSE), SSR/SST, SSE/SST, and Theil's U1 and U2 statistics. Table 6.6 depicts the error metrics, from which we can observe that for all the four datasets, ε-TSVR and TSVR models outperform conventional SVR and LS-SVR models in terms of RMSE and MAE. For dataset Anholt labeled as **Xx**, ε-TSVR accounts for maximum IOA, which indicates a good agreement between actual and

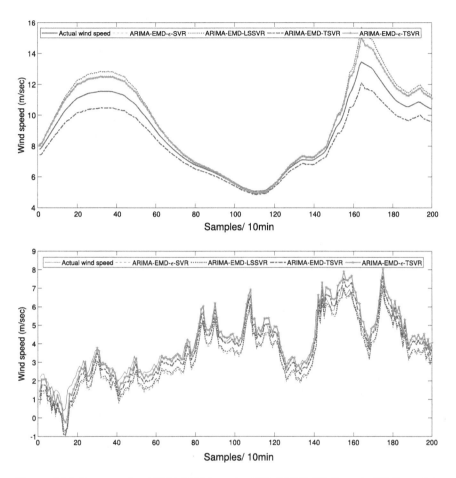

Figure 6.11 Forecasts for ARIMA-EMD-based ε-SVR, LSSVR, TSVR, and ε-TSVR models for datasets Anholt and Bishop & Clerks.

predicted values. In terms of RMSE, ε-TSVR is 27.22% better than the conventional SVR model. Similarly, for dataset Bishop & Clerks, ε-TSVR model outperforms conventional SVR, LS-SVR, and TSVR in terms of RMSE and MSE. However, it is worth noting that computationally, LS-SVR is the fastest among all regressors.

In this chapter, we have studied hybrid machine intelligent SVR models for short-term wind forecasting built on wavelet transform and empirical mode decomposition technique. A fourth-order Daubechies (db4) wavelet filter is chosen to carry out the wind speed time-series decomposition for four different wind farm sites. Among these regressors, the hybrid model

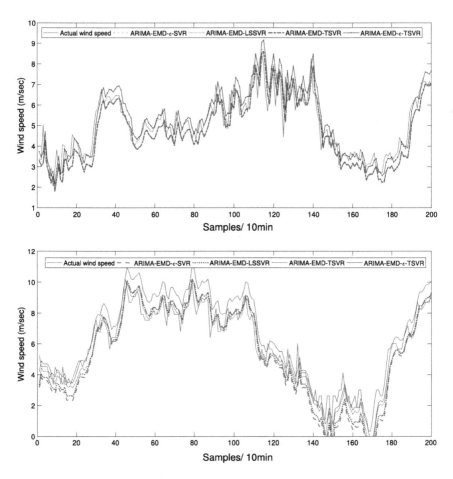

Figure 6.12 Forecasts for ARIMA-EMD-based ε-SVR, LSSVR, TSVR, and ε-TSVR models for dataset Blandford and Paxton.

based on wavelet transform and ARIMA-EMD, TSVR, and ε-TSVR prove to be better short-term forecast choices based on the performance indices for the four datasets. Computationally, LS-SVR takes the minimum time on CPU, and ε-TSVR takes less computation time than ε-SVR, owing to its smaller-sized optimization problem. The wind speed forecasting accuracy for all the hybrid models can be further improved by optimally selecting the SVR hyperparameters RBF bandwidth and regularization constants. Further, the effect of large dataset is examined for datasets C and G. Results reveal that ε-TSVR method is superior to TSVR, LS-SVR, and ε-SVR in terms of RMSE and MAE.

Table 6.6 Performance metrics of ARIMA-EMD-based ε-SVR, LS-SVR, TSVR, and ε-TSVR models.

Dataset	Model	RMSE	MSE	SSR/SST	SSE/SST
		IOA	U1	U2	CPU time
Xx	ε-SVR	0.8918	0.7953	1.584	0.1371
		0.9998	0.0460	0.9102	1.05007
	LSSVR	0.8577	0.7356	1.6008	0.1268
		0.9998	0.0444	0.9067	0.0889
	TSVR	0.7532	0.5673	0.7744	0.0978
		0.9998	0.0422	1.0939	0.5595
	ε-TSVR	0.6490	0.4212	1.4008	0.0726
		0.9999	0.0339	0.8895	0.2756
D	ε-SVR	0.8259	0.6821	1.619	0.3280
		0.9995	0.1056	0.4390	0.7385
	LS-SVR	0.7300	0.5329	1.538	0.2563
		0.9996	0.0922	0.3353	0.1220
	TSVR	0.5182	0.2685	1.5329	0.1291
		0.9998	0.0623	1.1374	0.6391
	ε-TSVR	0.4963	0.2463	1.5231	0.1184
		0.9998	0.0577	5.4156	0.93116
C	ε-SVR	0.6822	0.4653	1.3097	0.2354
		0.9996	0.0669	2.2648	0.8444
	LS-SVR	0.6854	0.4697	1.2095	0.2355
		0.9996	0.0668	4.8873	0.0876
	TSVR	0.6899	0.4759	1.2460	0.2366
		0.9996	0.0672	8.3949	0.5567
	ε-TSVR	0.6573	0.4320	1.3191	0.2148
		0.9997	0.0610	1.2391	0.5432
B	ε-SVR	6.556	42.98	6.926	7.259
		0.9970	0.0992	0.9804	0.8300
	LS-SVR	0.9765	0.9535	1.5738	0.1607
		0.9998	0.0777	0.5870	0.0965
	TSVR	0.7969	0.6350	1.2710	0.1070
		0.9998	0.0636	0.6254	0.5764
	ε-TSVR	0.9875	0.9751	1.5566	0.1647
		0.9998	0.0741	1.4054	0.6753

References

[1] H.S. Dhiman, D. Deb, J.M. Guerrero, Hybrid machine intelligent SVR variants for wind forecasting and ramp events, Renewable & Sustainable Energy Reviews 108 (2019) 369–379, https://doi.org/10.1016/j.rser.2019.04.002.

[2] E. Cadenas, W. Rivera, Wind speed forecasting in three different regions of Mexico, using a hybrid ARIMA–ANN model, Renewable Energy 35 (12) (2010) 2732–2738.

[3] D. Liu, D. Niu, H. Wang, L. Fan, Short-term wind speed forecasting using wavelet transform and support vector machines optimized by genetic algorithm, Renewable Energy 62 (2014) 592–597.

[4] C. Zhang, H. Wei, X. Zhao, T. Liu, K. Zhang, A Gaussian process regression based hybrid approach for short-term wind speed prediction, Energy Conversion and Management 126 (2016) 1084–1092.

[5] X. Mi, H. Liu, Y. Li, Wind speed forecasting method using wavelet, extreme learning machine and outlier correction algorithm, Energy Conversion and Management 151 (2017) 709–722.

[6] H. Li, J. Wang, H. Lu, Z. Guo, Research and application of a combined model based on variable weight for short term wind speed forecasting, Renewable Energy 116 (2018) 669–684.

[7] P. Jiang, Y. Wang, J. Wang, Short-term wind speed forecasting using a hybrid model, Energy 119 (2017) 561–577.

[8] R. Azimi, M. Ghofrani, M. Ghayekhloo, A hybrid wind power forecasting model based on data mining and wavelets analysis, Energy Conversion and Management 127 (2016) 208–225.

[9] Y. Jiang, G. Huang, X. Peng, Y. Li, Q. Yang, A novel wind speed prediction method: hybrid of correlation-aided DWT, LSSVM and GARCH, Journal of Wind Engineering and Industrial Aerodynamics 174 (2018) 28–38.

[10] H. Liu, Z. Duan, F. Han, Y. Li, Big multi-step wind speed forecasting model based on secondary decomposition, ensemble method and error correction algorithm, Energy Conversion and Management 156 (2018) 525–541.

[11] C. Tian, Y. Hao, J. Hu, A novel wind speed forecasting system based on hybrid data preprocessing and multi-objective optimization, Applied Energy 231 (2018) 301–319.

[12] J. Wang, W. Yang, P. Du, Y. Li, Research and application of a hybrid forecasting framework based on multi-objective optimization for electrical power system, Energy 148 (2018) 59–78.

[13] P. Du, J. Wang, W. Yang, T. Niu, Multi-step ahead forecasting in electrical power system using a hybrid forecasting system, Renewable Energy 122 (2018) 533–550.

[14] M. Silva, M. Oleskovicz, D. Coury, A hybrid fault locator for three-terminal lines based on wavelet transforms, Electric Power Systems Research 78 (11) (2008) 1980–1988.

[15] T. Qian, M.I. Vai, Y. Xu (Eds.), Wavelet Analysis and Applications, Birkhäuser, Basel, 2007.

[16] Y. Hao, C. Tian, The study and application of a novel hybrid system for air quality early-warning, Applied Soft Computing 74 (2019) 729–746.

[17] Weather history download Macarthur – meteoblue, https://www.meteoblue.com, 2019. (Accessed 9 March 2019).

[18] Merra – www.soda-pro.com, http://www.soda-pro.com, 2019. (Accessed 11 March 2019).

[19] F.X. Diebold, Comparing predictive accuracy, twenty years later: a personal perspective on the use and abuse of Diebold–Mariano tests, Journal of Business & Economic Statistics 33 (1) (2015) 1.

[20] G.-F. Fan, S. Qing, H. Wang, W.-C. Hong, H.-J. Li, Support vector regression model based on empirical mode decomposition and auto regression for electric load forecasting, Energies 6 (4) (2013) 1887–1901.

[21] W. yang Duan, L. min Huang, Y. Han, Y. hui Zhang, S. Huang, A hybrid AR-EMD-SVR model for the short-term prediction of nonlinear and non-stationary ship motion, Journal of Zhejiang University. Science A 16 (7) (2015) 562–576.

[22] W. Duan, Y. Han, L. Huang, B. Zhao, M. Wang, A hybrid EMD-SVR model for the short-term prediction of significant wave height, Ocean Engineering 124 (2016) 54–73.

[23] Y. Xiang, L. Gou, L. He, S. Xia, W. Wang, A SVR–ANN combined model based on ensemble EMD for rainfall prediction, Applied Soft Computing 73 (2018) 874–883.

[24] C.-J. Yu, Y.-Y. He, T.-F. Quan, Frequency spectrum prediction method based on EMD and SVR, in: 2008 Eighth International Conference on Intelligent Systems Design and Applications, IEEE, 2008.

[25] H.-L. Yang, H.-C. Lin, An integrated model combined ARIMA, EMD with SVR for stock indices forecasting, International Journal on Artificial Intelligence Tools 25 (02) (2016) 1650005.

CHAPTER 7

Ramp prediction in wind farms

Globally, wind energy has lessened the burden on conventional fossil fuel-based power generation. Wind resource assessment for onshore and offshore wind farms aids in accurate forecasting and analyzing the nature of ramp events. Ramp events are scenarios in wind farms where the wind speed changes over a small amount of time leading to large power change. From an industrial point of view, a large ramp event in a short time duration is likely to cause damage to the wind farm connected to the utility grid. In this chapter, ramp events are predicted using hybrid machine intelligent techniques such as support vector regression (SVR) and its variants, random forest regression, and gradient boosted machines for onshore and offshore wind farm sites. A wavelet transform-based signal processing technique is used to extract features from wind speed. Results reveal that SVR-based prediction models give the best forecasting performance. In addition, gradient boosted machines (GBM) predict ramp events closer to the twin support vector regression (TSVR) model. Furthermore, the randomness in ramp power is evaluated for onshore and offshore wind farms by calculating the log energy entropy of features obtained from wavelet decomposition and empirical mode decomposition.

7.1 Ramp events in scientific and engineering activities

As has been seen so far in this book, wind forms basis of many daily engineering applications. Primarily, it is tapped from atmosphere for generating electric power for commercial, residential, and industrial purposes [1]. Statistics have revealed that the wind power is a significant backbone in the power portfolio of many developing and developed countries [2]. The power generated from wind is efficient and has negligible carbon footprints, thus making it one of the cheapest sources of energy [3]. Wind turbines are often erected near agricultural lands and allow farmers to continue their farming practice as the area occupied by a single wind turbine is very small as compared to the surrounding available land. However, a major challenge with wind energy is its intermittency, which makes its reliability a serious challenge [4]. Through the advanced forecasting algorithms it is possible to predict the wind speed up to a certain level of accuracy. Further, siting a

Supervised Machine Learning in Wind Forecasting and Ramp Event Prediction
https://doi.org/10.1016/B978-0-12-821353-7.00018-1

proper wind farm location plays an important role in shaping the sustainability of grid integration of wind energy. Though there are several positive impacts of wind energy in terms of power generation and pollution-free source, it has also raised ecological concerns in terms of bird killings caused by turbine rotors and high noise levels [5]. Desholm [6] has ranked the bird species in terms of their sensitivity to bird-turbine collisions for Nysted wind farm in Denmark. A recent study in China based on learning curve analysis has stated that the wind farm performance has benefited largely from the experience of turbine manufacturers and wind farm developers, indicating an indirect financial impact of wind resource [7].

Wind speed also influences many commercial practices like agriculture, crop forecasting, flood forecasting, earthquake forecasting, and building energy demand forecasting [8]. The intermittent nature of wind speed can also lead to uprooting of essential crop under cultivation and thus reduce the yield. Martins et al. [9] have carried out a case study to forecast the yield of maize crop in Brazilian semiarid region, where various meteorological parameters like maximum and minimum temperatures, rainfall, air pressure, and wind speed at height of 10 m are chosen to forecast crop yield for drought management. Apart from the wind speed, several other meteorological variables like the ambient air temperature, soil temperature, and relative humidity play a significant role in agricultural forecasts [10]. Important agricultural decision making processes like harvesting, field preparation, sowing, and irrigation are dependent on weather variables, one of them being the wind speed.

Agriculture is a primary sector in many developing countries, and its contribution to a country's gross domestic product (GDP) is generally high [11]. Globally, it has been reported that climate change impacts various agricultural practices in terms of nutrient deficiency in crops, severe drought conditions in semiarid and arid regions and undesirable eutrophication thus depleting oxygen profile in water bodies [12]. Ullrich et al. [13] have studied the impact of land management practices on water quality and quantity by using the soil and water assessment tool (SWAT). Changing climatic conditions in terms of rapid and random variation in ambient temperature, wind speed, relative humidity have caused changes in crop yield of wheat, barley, and maize in Banas river basin in Rajasthan, India [14]. To assess the model accuracy, RMSE values were calculated and were found to be 11.99, 16.15, and 19.13 for wheat, barley, and maize, respectively. Robichaud et al. [15] have carried out wind tunnel experiments to evaluate the sediment losses caused due to extreme wind speed conditions (6,11 and 18 m/s). The sediment losses were reported to be ranging from 6.5 g/m² to 6258 g/m² for low to high wind speeds, respectively. Li et al. [16] have explored the effect of particulate matter (PM2.5) emissions on the agricultural growth of China in 2001–2010. Since the PM2.5 emissions are affected by the wind speed profile of a particular region, there has been an direct negative correlation between them.

Wind speed influences the soil erosion in agricultural lands, thus endangering the fertility and yield of crops. Sirjani et al. [17] have discussed the soil erosion phenomenon in a wind tunnel experiment for semiarid and arid regions for different soil properties like mean weight diameter, soil clay, and moisture content. A case study carried out by Wiggs et al. [18] demonstrates how changes in wind speed affect soil erosion and erodibility in west-central region of South Africa which is characterized by low rainfall. The measurements are taken for wind speed using an array of four cup anemometers placed at heights of 0.5 m, 0.9 m, 1.5 m, and 2.3 m, and magnitude of soil erosion is measured by surface mounted saltation flux impact recorder. The experiment is divided into seven event categories starting from August 7, 2007, to November 13, 2007. The soil erosion event is considered to occur whenever the impact wind speed increased or became equal to a threshold value. Experimental results indicate that soil erosion occurs majorly during increased average wind speed events, particularly, in the range of 3 m/s to 15 m/s. Rapid changes in wind speed also cause high-rise buildings and low-rise homes near coastal areas to get uprooted. However, the surface winds cause most of the damage, and their variability is highly dependent on the nature of the surface. The nature of the land is characterized by surface roughness length z_0, which differs from surface to surface.

The wind velocity also affects the height of the flood wave generated at the reservoir surface. An empirical relationship put forward by D.A. Molitor between height of flood wave z_d in meters and wind velocity in km/h given by [19] is expressed as

$$z_d = 0.032\sqrt{UF} + 0.75 - 0.27F^{1/4}, \tag{7.1}$$

where U is the wind velocity (km/h), and F is the fetch (km).

A case study for the effect of wind speed variation on the reservoir wall at two sites (Vadodara and Surat) in Gujarat, India, is studied in the next section. Further, Wang et al. [20] have carried out experimental study related to the effect of different wind speeds on cotton combustion and have reported a significant combustion rate in the presence of higher wind speeds. In terms of air quality the magnitude of the particulate matter in atmosphere is greatly influenced by the variability in wind speed. Pushpawela et al. [21] have studied the effect of wind speed on new particle formation in the subtropical urban environment in the city of Brisbane, Australia.

The new particle formation (NPF) events are recorded at the Queensland University of Technology in Brisbane with neutral cluster and air ion spectrometer (NAIS) mounted at the sixth floor of the building. It was reported that the NPF occurrence was triggered for more polluted cities (such as Beijing, China, and Po Valley in Italy) with higher wind speeds in the atmosphere predominately during morning hours from 8 AM to 9 AM. The impact of wind speed and ramp event in different scenarios is studied next in diverse case studies.

7.1.1 Reservoir wall: a case study from Valsad, Gujarat

Reservoirs are often found in use to store rainfall water and to mobilize it during the periods of scanty rainfall. Construction of such a storage mechanism requires a dam to be built on the natural course of river flow, referred as an impounding dam [22]. Wind speed flowing over the water surface of reservoirs often generate waves, which cumulatively impinge on the reservoir wall and exert force. Wave height generated at the reservoir surface can be determined empirically using Molitor's relationship between the wind velocity U and wave height z_d as stated in (7.1).

A case study is undertaken for Chasmandva reservoir located near the river Tan, a tributary of river Auranga in Valsad district of Gujarat (20°37′02″ N and 73°22′36″ E) with a fetch length of $F = 4.38$ km.

Fig. 7.1 illustrates the wind speed, wave height, and wave force on the reservoir wall for the Chasmandva reservoir, Valsad, Gujarat. The variability in wind speed is seen during March–May, where the average wind speed changes significantly from 5 km/h to 15 km/h. During this change in wind speed, the average bending moment on reservoir wall rises significantly from 235.48 kN·m to 344.55 kN·m, leading to an increase of 46.32% as determined by (7.4).

The monthly average wind speed data is collected at height of 10 m from Indian Meteorological Department (IMD) observatory located at Surat, Gujarat, for the years 1998–2007. Next, we extrapolate the wind speed using log-law at a reservoir water surface at a height determined by subtracting the full reservoir level (FRL = 214 m) and deepest bed level

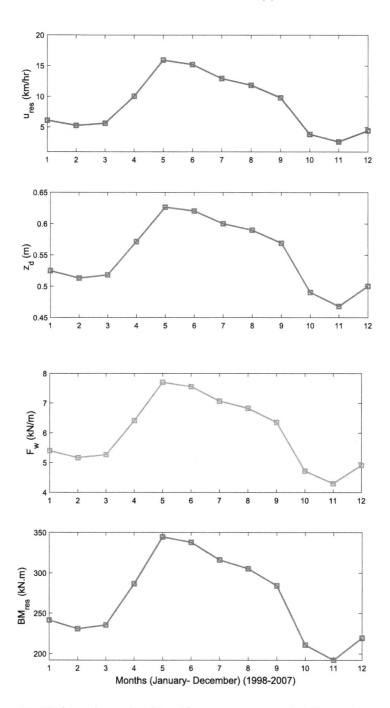

Figure 7.1 Wind speed, wave height and force on reservoir wall at Chasmandva.

(DBL $= 169.5$ m). The wind speed log-law stated by [23] is given as

$$\frac{u_{res}}{u_{10}} = \frac{\log\left(\frac{z_{res}}{z_0}\right)}{\log\left(\frac{z_{10}}{z_0}\right)}, \tag{7.2}$$

where u_{res} and u_{10} are the wind speed (in km/h) at reservoir height and reference height, respectively, and z_{res}, z_{10}, and z_0 are the reservoir height, reference height, and surface roughness length, respectively. The wave height due to wind speed at water surface can be found using (7.1).

The waves so generated due to wind speed at the water surface cause a force and bending moment to be exerted on the dam wall given as

$$F_w = 2\gamma_w z_d^2, \tag{7.3}$$

$$BM_{res} = F_w\left(D_{res} + \frac{3}{8}z_d\right), \tag{7.4}$$

where γ_w is the specific gravity of water, z_d is the wave height, and D_{res} is the depth of still water in reservoir. High ramp rates of wind speed cause significant wave force on the reservoir wall and may put the stability of hydraulic structure in jeopardy.

7.1.2 Forest risk management: a case study

Forests cover 30.825% of Earth's land surface and appropriate management strategies for averting the risk caused by the random nature of wind speed form an essential component in arboriculture studies [24]. However, the nature of wind profile in the forest area is complicated due to complex structure of vegetation in terms of leaf size, shape, and terrain [25]. The wind speeds flowing in the forest region not only cause risks associated with trees and shrubs but also cause soil erosion and transportation of crop insects from one place to another [26]. The damages caused by wind speed are often related to the heterogeneous nature of canopy and hilly topography [27]. When wind flows through the forest, it experiences a reduction in its magnitude due to resistance offered by obstructions such as stem, leaves, and branches. The major concern among the forest management associations globally is the hazard of forest fire due to rapid changes in wind speed.

A case study for analyzing the impact of ramp wind speeds on forests is studied for Montesinho Natural Park, Portugal. The park is rich in flora and fauna and has several types of trees such as Pyranean oak, chestnut oaks, willows, and holm oaks. However, northern Portugal is highly abundant in Pyranean oaks with a height from 20 m to 25 m. The speed data is collected every hour at meteorological tower in central Montesinho park. Wind speed measurement is carried out using a set of cup anemometers placed at height of 10 m above the ground. Several other weather variables such as temperature (in °C), relative humidity (%), and area burnt (in ha) are also collected. The change in wind speed is calculated for the selected duration and is given as

$$\Delta u_{for} = u_{for}(t + \Delta t) - u_{for}(t), \qquad (7.5)$$

where u_{for} (in km/h) is the wind speed in the forest canopy. Since the forest fires are erratic throughout the year, the area burnt time series is transformed using logarithmic transformation and is plotted for its frequency distribution. Fig. 7.2 illustrates the frequency distribution for ramp speed and area burnt. The ramp wind speed follows a normal distribution indicated by red solid line, and random changes in its magnitude lead to increased risk of forest fires as indicated by an upward increasing trend with rapid changes in wind speed.

7.1.3 Strawberry cultivation case study, Monterey, CA

Strawberry is an essential fruit crop cultivated widely across the regions of California and Florida in United States and Spain, with California being the leader in production. Weather dependency of strawberry yields has led to forecast of important variables like ambient temperature, relative humidity, wind speed, and soil temperature. Climate forecasts help the strawberry farm operators to plan their harvest ahead of time. Water management practices also influence the yield of strawberries. Another important issue with strawberry cultivation is the presence of insects/pests that reduce the effective yield of strawberries with western tarnished plant bug and two-spotted spider mite being major culprits.

We study the impact of wind speed variability along with soil temperature on the yield of strawberries for California state. The Monterey city of California is located on the central coast and accounts for 32.9% of California's strawberry production. The wind speed (in m/s) data along with soil temperature (°C) and yield (in kg/ha) are obtained from California Irrigation Management Information System (CIMIS) for one week duration from September 5, 2018, to September 11, 2018. Correlations among wind speed, soil temperature, ramp wind speed (change in wind speed), and strawberry yield were calculated. Pearson's correlation coefficient between

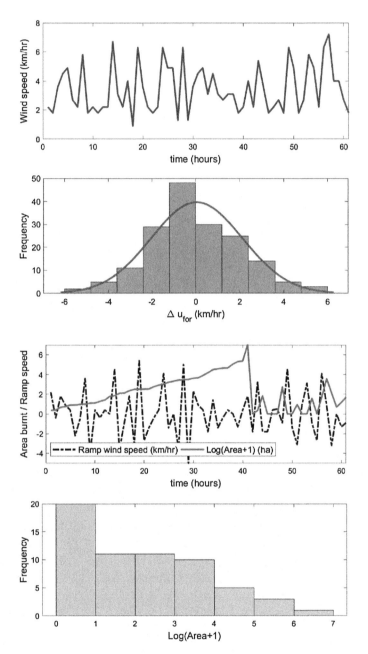

Figure 7.2 Frequency distribution for ramp wind speed, forest area burnt, wind speed pattern and variation of area burnt and ramp wind speed, Montesinho Natural Park, Portugal.

two variables x and y is given as

$$\rho_{xy} = \frac{n\sum(xy) - \sum x \sum y}{\sqrt{(n\sum x^2 - (\sum x)^2)(n\sum y^2 - (\sum y)^2)}}. \tag{7.6}$$

The following matrix shows the coefficients of Pearson's correlation:

$$
Z_{corr} =
\begin{array}{cccc}
\text{Soil temp} & \text{Yield} & \text{Wind speed} & \text{Ramp speed} \\
\left[\begin{array}{cccc}
1 & -0.277 & -0.221 & 0.098 \\
-0.277 & 1 & 0.685 & 0.798 \\
-0.221 & 0.685 & 1 & 0.354 \\
0.098 & 0.798 & 0.354 & 1
\end{array}\right] &
\begin{array}{l}
\text{Soil temp} \\
\text{Yeild} \\
\text{Wind speed} \\
\text{Ramp speed}
\end{array}
\end{array}
$$

where Z_{corr} is Pearson's correlation matrix indicating the correlations among wind speed, ramp wind speed, soil temperature, and strawberry yield. From the matrix we can see that the correlation between wind speed and yield is 0.685, indicating a strong positive correlation. Further, the correlation between ramp wind speed (Δu) and yield is 0.798, depicting that variability in wind speed influences the yield of strawberry. The yield of strawberry crop is also affected by the concentration of PM2.5 pollutant in atmosphere. The correlation coefficient between daily average PM2.5 levels and strawberry yield is given as

$$
ZH_{corr} =
\begin{array}{cc}
\text{Yield} & \text{PM2.5} \\
\left[\begin{array}{cc}
1 & 0.513 \\
0.513 & 1
\end{array}\right] &
\begin{array}{l}
\text{Yeild} \\
\text{PM2.5}
\end{array}
\end{array}.
$$

7.1.4 Air quality: a case study for Bogota, Colombia

Airborne particles and gases in the atmosphere not only deplete the air quality but also hamper the balance of the ecosystem in terms of health hazards and skewed food chain processes. Major factors that contribute to the depleting air quality around the world are industrial and vehicular emissions. Ambient temperature, wind speed, relative humidity, wind gusts, and cloud cover affect the air quality index (AQI) for a specific region. AQI-related studies have been carried out that typically relate the influence of wind speed on the particulate matter (PM 2.5 and PM10) concentration in the atmosphere [28]. According to a study carried out by the World Health Organization (WHO), the severity of respiratory diseases is strongly correlated to SO_2 pollutants.

We study the impact of wind speed and ramp wind speed on the air quality index (AQI) and particulate matter concentration (PM2.5) for the city of Bogota, Colombia. The wind speed (in m/s) data along with AQI and PM2.5 concentration (in $\mu g/cm^3$) is collected from Environmental Protection Agency (US) for a duration of 14 days from September 1, 2018, to September 14, 2018 [29]. Pearson's correlation coefficient between PM2.5 concentration, wind speed, and wind ramp speed is evaluated, and Pearson's correlation matrix is given as

$$H_{corr} = \begin{matrix} & \text{Wind speed} & \text{Wind gusts} & \text{PM2.5} & \text{Wind ramp} & \\ & \begin{bmatrix} 1 & 0.85 & -0.319 & 0.193 \\ 0.85 & 1 & -0.322 & 0.14 \\ -0.319 & -0.322 & 1 & -0.053 \\ 0.193 & 0.14 & -0.053 & 1 \end{bmatrix} & \begin{matrix} \text{Wind speed} \\ \text{Wind gusts} \\ \text{PM2.5} \\ \text{Wind ramp} \end{matrix} \end{matrix}$$

There exists a weak negative correlation between PM2.5 and wind speed. This particularly suggests that the increase in wind speed reduces the PM2.5 concentration levels and vice versa. The random and rapid changes in wind speed and direction also have significant impact on the air quality index suggesting its benefit in regions with heavy industrial emissions. Fig. 7.3 illustrates the wind speed, AQI, PM 2.5 concentration levels, and ramp wind speed for Bogota, Colombia. See Figs. 7.4 and 7.5.

Regression analysis is studied to determine the effect of wind ramps on the PM2.5 concentration levels. The regression is performed on a one month hourly data for the city of Bogota, Colombia, and is carried out in SPSS environment. The inputs for the regression model are the wind speed measured at a height of 10 m and 80 m, the air quality index (AQI), the wind gusts measured at surface, and the ramp speed at heights of 10 m and 80 m. Two cases are analyzed, without ramp speed (Case A) and with ramp speed (Case B). Table 7.1 depicts various performance metrics.

Next, we carry out the regression analysis for PM2.5 levels for the city of Abu Dhabi, UAE. The wind speed data along with ramp wind speed and air quality index are used as independent variables for predicting PM2.5 levels. The hourly data is collected from Environmental Protection Agency, USA, for the duration from September 3, 2018, to September 15, 2018. The regression is based on the autoregressive integrated moving average (ARIMA) method with p, d, and q being the orders for the autoregression, differencing, and moving average models. A standard ARIMA(p, d, q)

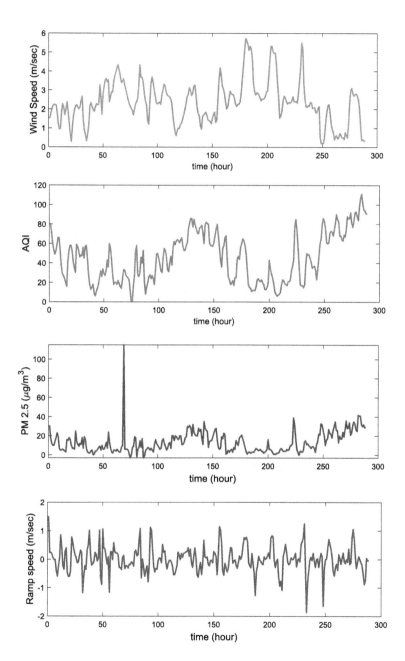

Figure 7.3 Wind speed, AQI, PM 2.5 levels and ramp wind speed for Bogota.

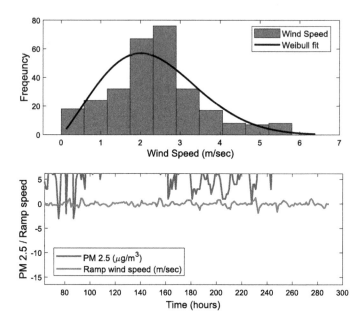

Figure 7.4 Wind speed, wind gusts frequency distribution and variation of AQI and PM2.5 levels with ramp wind speed for Bogota, Colombia.

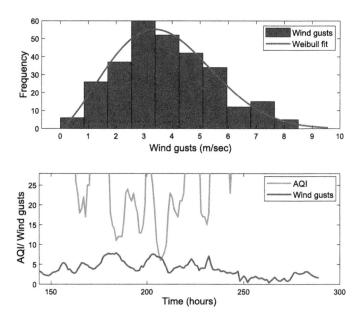

Figure 7.5 Wind speed, wind gusts frequency distribution and variation of AQI and PM2.5 levels with ramp wind speed for Bogota, Colombia.

Table 7.1 Regression analysis for PM2.5 levels in Bogota, Colombia.

Performance metric	Case A	Case B
RMSE	0.8632	0.8212
R²	0.9213	0.9498

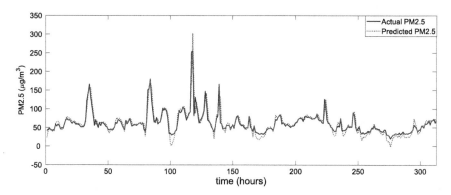

Figure 7.6 Forecasting PM2.5 level for Abu Dhabi, UAE.

model is given as

$$\hat{y} = \mu + \phi_1 y_{t-1} + \phi_2 y_{t-2} + \cdots + \phi_p y_{t-p} - \theta_1 e_1 - \cdots - \theta_q e_{t-q}, \qquad (7.7)$$

where $\hat{y}$ is the predicted value of variable of interest, ϕ_i $(i = 1, 2, \ldots, p)$ are the coefficients for autoregressive process, θ_i $(i = 1, 2, \ldots, q)$ are the coefficients for moving average process, and e_{t-q} is the error term with qth order of moving average model. ARIMA(1,0,1) is used for predicting the future values of PM2.5 levels. The order (p, d, q) can be found out by calculating the autocorrelation of PM2.5 time series. The prediction accuracy is measured by calculating the root mean squared error (RMSE) between actual values and predicted values and is given as

$$RMSE = \sqrt{\frac{1}{N} \sum_{i=1}^{N} (\hat{y}_i - y_i)^2}, \qquad (7.8)$$

where $\hat{y}_i$ and y_i are the predicted and actual values for PM2.5 concentration levels for N observations. The RMSE calculated is found to be 12.585. Fig. 7.6 illustrates the predicted and actual values for PM2.5 concentration levels.

So far, in this chapter, we have analyzed the impact of wind speed and its variability on different scientific and engineering aspects. Next, we discuss four case studies for studying the impact of wind ramps. The random and rapid changes in wind speed are termed as a ramp and have several catastrophic impacts. First, we studied the case study for a reservoir wall at Chasmandva district in Valsad, Gujarat. The wave height generated at the water surface due to wind speed is calculated, and its corresponding force and bending moment on the reservoir wall are determined, and it is found that from March to May the bending moment is increased by 46.32% due to the mean wind speed change from 5 km/h to 15 km/h. Then we study the impact of wind speed on the forest risk management for a forest canopy in Portugal, where the wind ramps cause forest fires to spread rapidly with rapid changes in wind speed. The wind speed impact is then assessed for strawberry cultivation for a farm in Monterey in California, where the correlation between daily yield and ramp speed is found to be 0.798, thus indicating wind ramps as favorable scenario for strawberry yield. Then we analyzed the influence of wind speed in the air quality of the city of Bogota, Colombia, and Abu Dhabi, UAE. Pearson's correlation coefficient is calculated between PM2.5 concentration level and wind ramp speed and was found to be -0.053, indicating improvement of air quality conditions with increasing wind ramp speed magnitude.

7.2 Ramp events in wind farms

Intermittent wind power due to sudden wind speed variations is critical for grid connected power plants and causes extreme situations like low system reliability and high operational costs. A wind power ramp event is defined as the rate of change in wind power generated over a short time period exceeding a predefined threshold value (normally 50%) [30]. According to [31], a power ramp event is said to occur if the change in power signal $|P(t + \Delta t) - P(t)|$ is greater than a said threshold, ΔP_{ramp}. The intermittent nature of wind speed requires installation of batteries to tackle peak demands. However, rapid charging and discharging of batteries degrade battery life [32]. For analysis of power ramp up or down events, setting the threshold power is needed. For a given wind turbine, let the ramp threshold power be $r\%$ of the nominal wind power. Then we can define two different ramp thresholds:

$$\Delta P_{ramp} = \begin{cases} + r\% \ \text{of} \ P_{nominal} = P_{th}^{u}, \\ - r\% \ \text{of} \ P_{nominal} = P_{th}^{l}, \end{cases} \tag{7.9}$$

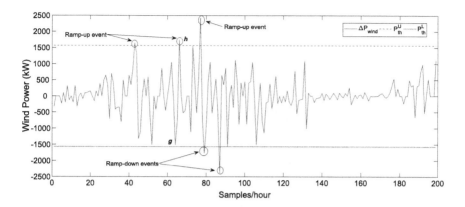

Figure 7.7 Schematic representation of wind power ramp events.

where P_{th}^u and P_{th}^l are the upper and lower ramp thresholds for ramp-up and ramp-down events, respectively, in a given period of time. We compare hybrid forecasting methods based on wavelet transform and ε-SVR, LS-SVR, TSVR, and ε-TSVR during power ramp events.

7.2.1 Ramp event error analysis for ε-SVR and LS-SVR

Consider a power ramp-up event at points g and h as shown in Fig. 7.7.

Let the wind powers at points g and h be P_g and P_h, and let $\Delta P_{gh} = P_h - P_g$ denote the change in wind power over a time interval ΔT. According to ε-SVR and LS-SVR, the forecasted values are

$$\hat{P}_{g1} = (\alpha_g - \alpha_g^*)k(x, x_g) + b, \quad \hat{P}_{h1} = (\alpha_h - \alpha_h^*)k(x, x_h) + b, \quad (7.10)$$

$$\hat{P}_{g2} = \eta_g k(x, x_g) + b_1, \qquad \hat{P}_{h2} = \eta_h k(x, x_h) + b_1, \quad (7.11)$$

at ramp points g and h such that α_g and η_g are the Lagrangian multipliers, b and b_1 are the bias terms, and $\hat{P}_{g1}$, $\hat{P}_{h1}$ and $\hat{P}_{g2}$, $\hat{P}_{h2}$ are the predicted values based on ε-SVR and LS-SVR models, respectively. The error in the predicted value $\hat{P}_{g1}$, $\hat{P}_{h1}$ and actual value P_g, P_h is given as

$$e_{SVR} = \hat{P}_{h1} - P_h - \hat{P}_{g1} - P_g, \quad e_{LS-SVR} = \hat{P}_{h2} - P_h - \hat{P}_{g2} - P_g, \quad (7.12)$$

where e_{SVR} and e_{LS-SVR} are the errors based on ε-SVR and LS-SVR models, respectively. Since the ramp event occurs between two set points, the error takes up the form as described by (7.12). If LS-SVR outperforms

ε-SVR, then we have $e_{LS-SVR} < e_{SVR}$, that is,

$$\hat{P}_{h1} - P_h - \hat{P}_{g1} + P_g > \hat{P}_{h2} - P_h - \hat{P}_{g2} + P_g. \tag{7.13}$$

Let us denote $\beta_h = \alpha_h - \alpha_h^*$ and $\beta_g = \alpha_g - \alpha_g^*$, and simplifying (7.13), we get

$$\beta_h k(x, x_h) + b - \beta_g k(x, x_g) - b > \eta_h k(x, x_h) + b_1 - \eta_g k(x, x_g) - b_1. \tag{7.14}$$

Since the kernel matrix elements $k(x, x_i)$ are equal for ε-SVR and LS-SVR, the equation can be further simplified as

$$k(x, x_h)\left(\beta_h - \eta_h\right) - k(x, x_g)\left(\beta_g - \eta_g\right) > 0. \tag{7.15}$$

Thus, if (7.15) is satisfied, then LS-SVR will outperform ε-SVR during ramp events.

7.2.2 Ramp event error analysis for TSVR and ε-TSVR

Similarly, TSVR and ε-TSVR can be compared based on the same approach. Let e_{TSVR} and $e_{\varepsilon-TSVR}$ denote the errors in the wind ramp power between points g and h based on TSVR and ε-TSVR, respectively. We have

$$\hat{P}_{g3} = \frac{1}{2}(w_1 + w_2)k(x, x_g) + \frac{1}{2}(b_1 + b_2), \tag{7.16}$$

$$\hat{P}_{h3} = \frac{1}{2}(w_1 + w_2)k(x, x_h) + \frac{1}{2}(b_1 + b_2), \tag{7.17}$$

$$\hat{P}_{g4} = \frac{1}{2}(u_1 + u_2)k(x, x_g) + \frac{1}{2}(b_3 + b_4), \tag{7.18}$$

$$\hat{P}_{h4} = \frac{1}{2}(u_1 + u_2)k(x, x_h) + \frac{1}{2}(b_3 + b_4), \tag{7.19}$$

where b_1, b_2 and b_3, b_4 are bias terms, and $\hat{P}_{g3}$, $\hat{P}_{h3}$ and $\hat{P}_{g4}$, $\hat{P}_{h4}$ are the predicted values of wind power using TSVR and ε-TSVR, respectively. The forecasted ramp power $\hat{\Delta} P_{gh}$ is then compared, and the errors in ΔP_{gh} for TSVR and ε-TSVR are

$$e_{TSVR} = \hat{P}_{h3} - P_h - \hat{P}_{g3} + P_g, \quad e_{\varepsilon-TSVR} = \hat{P}_{h4} - P_h - \hat{P}_{g4} + P_g. \tag{7.20}$$

Comparing the ramp power errors, if $e_{TSVR} > e_{\varepsilon-TSVR}$, then we get

$$\hat{P}_{h3} - P_h - \hat{P}_{g3} + P_g > \hat{P}_{h4} - P_h - \hat{P}_{g4} + P_g. \tag{7.21}$$

Let us define $\frac{1}{2}(w_1 + w_2) = \hat{w}$ and $\frac{1}{2}(u_1 + u_2) = \hat{u}$, and by simplifying (7.21) we get

$$\hat{w}\Big(k(x, x_h) - k(x, x_g)\Big) > \hat{u}\Big(k(x, x_h) - k(x, x_g)\Big). \tag{7.22}$$

When (7.22) is satisfied, ε-TSVR outperforms TSVR during ramp events between points g and h.

7.2.3 Case study for ramp event analysis

In our study, we choose five wind farms namely, Sotavento (Spain), Paxton and Blandford (MA), Beresford (South Dakota) [33], and AGL Macarthur wind farm, Victoria, Australia, to analyze the wind power ramp events. The threshold ramp power is chosen as 15% of nominal power (P_{nom}). Wind turbines (Vestas V112) from the Danish manufacturer Vestas with rated speed 12 m/s are selected to study the wind power ramp event. Two ramp events, power ramp-up and power ramp down events are studied. The nominal wind power of the given wind turbine is 3.6496 MW. The threshold limit for ramp power events is chosen as 15% of the nominal power. Fig. 7.8 shows the frequency distribution of change in wind power (ΔP_{wind}) in successive dispatch windows for different hub heights.

The wind speed data at hub height of 10 meters is transformed at a hub height of 40 m using the wind profile power law [34] given as

$$\frac{u_h}{u_r} = \left(\frac{z_h}{z_r}\right)^\alpha, \tag{7.23}$$

where u_h and u_r are the wind speeds (in m/s) at desired hub height and reference hub height, z_h and z_r are the hub heights (in m) at desired level and reference level, respectively, and $\alpha = 1/7$ is an empirically calculated constant dependent on atmospheric conditions [35]. Table 7.2 shows the absolute error (AE) values computed for wind power ramp-up and ramp-down events for different wind farm sites. During ramp-up events for all the wind sites, ε-TSVR performs better than TSVR, LS-SVR, and classical ε-SVR.

From Fig. 7.8 we see that the probability of wind power ramp event increases when wind speed is recorded at any hub height above the ground. The number of wind power ramp events for five wind farm sites are illustrated in Fig. 7.9.

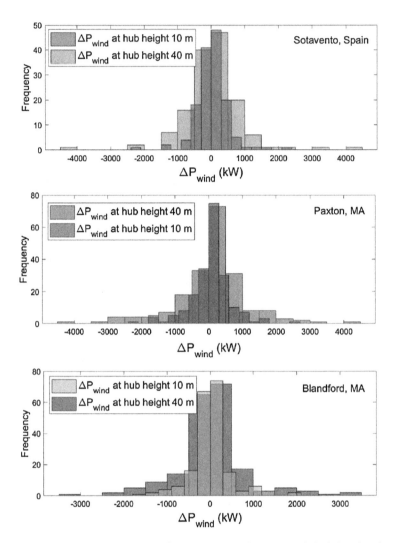

Figure 7.8 Frequency distribution of change in wind power with hub heights for Sotavento, Paxton and Blandford.

7.3 Ramp event analysis for onshore and offshore wind farms

Recent advances in renewable energy sector have streamlined the power generation portfolio of many developing countries. With wind energy being the driving force, both onshore and offshore technologies have attracted big investments globally [36]. Onshore wind farms have an advantage of

Table 7.2 Performance metric (AE) during wind power ramp events.

Wind Farm	Model	Wind power ramp event	
		Ramp-up	Ramp-down
Sotavento, Spain	ε-SVR	0.7245	0.7626
	LS-SVR	0.4587	0.5303
	TSVR	0.3600	0.4191
	ε-TSVR	0.1414	0.2019
Paxton, MA	ε-SVR	0.0454	0.0174
	LS-SVR	0.0347	0.0133
	TSVR	0.0118	0.0055
	ε-TSVR	0.0018	0.0104
Blandford, MA	ε-SVR	0.0574	0.8058
	LS-SVR	0.0350	0.1650
	TSVR	0.0237	0.1454
	ε-TSVR	0.0161	0.1211
Beresford, South Dakota	ε-SVR	0.0657	0.1074
	LS-SVR	0.0500	0.1058
	TSVR	0.0074	0.0025
	ε-TSVR	0.0067	0.0021
AGL Macarthur, Victoria, Australia	ε-SVR	0.2540	0.5497
	LS-SVR	0.1221	0.3624
	TSVR	0.0669	0.1877
	ε-TSVR	0.0570	0.2821

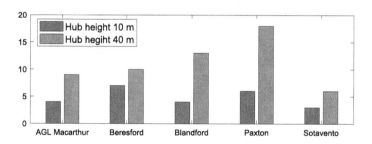

Figure 7.9 Frequency of wind power ramp events with different hub heights.

proximity to the utility grid, whereas offshore wind farms need long transmission cables to transmit power from the sea to grid. However, with a strong wind field in the offshore areas compared to onshore, which significantly compensates for nonproximity to the grid, the offshore wind farm installations have risen drastically. Further, with lesser turbulence and more uniform wind speed, the lifetime of wind turbines in offshore scenario increases [37]. Offshore wind farms have a disadvantage of higher installation and engineering costs accompanied by the wear and tear of generator and mechanical equipments. On the other hand, onshore wind farms face the challenge in their acceptance from limited land availability, visual intrusion, and damage to wildlife. The defining factor of operation between onshore and offshore wind farms is the surface roughness length. It is observed that the larger the surface length, the higher the reduction in wind velocity, and theoretically the wind speed at the ground is zero [38].

Wind energy potential in onshore and offshore areas has been an active research field. The installation activity, particularly, in the offshore areas, is challenged by the transportation costs owing to large blades and support structures to be placed in deep sea water. Another issue faced by the offshore wind turbines is the problem of uncontrolled vibrations arising from the combination of wind and wave force [39]. With a higher wind speed in the sea, the probability of occurrence of ramp events increases. Wind power growth in European countries has led to detailed study of ramp events, particularly, categorized as ramp-up and ramp-down events, each characterized by sudden change in wind speed in a short interval of time. Mathematically, a wind power ramp event is described as

$$\Delta P_w^{Ramp} = P_w(t + \Delta t) - P_w(t), \tag{7.24}$$

where ΔP_w^{Ramp} is the ramp wind power expressed as the arithmetic difference of two consecutive time intervals.

The occurrence of a wind power ramp event is ascertained by a threshold value of the total wind farm power. Wind power forecasting is an essential market procedure to regulate the electricity markets where the farm operators hold an advantage to derive profit from their generation given an accurate wind forecasting strategy. Predicting these ramp events can suitably prevent intermittent grid failures, and for that purpose, a proper characterization of ramp events is imperative. For example, in Horns Rev wind farm, large variation in wind power is seen in a short duration [40]. In a study carried out by Gjerstad the variation in wind power is triggered by atmospheric properties of boundary layer [41].

The atmospheric factor that influences the wind speed variations is the mean lapse rate $\partial T/\partial z$, that is, rate of change of air temperature T with height z. In case of offshore wind farms connected to German power grid, within 8 h, the wind power transfer between converters increased from

4 GW to 19 GW as a result of a sudden change in wind speed [42]. Further, Nissen studies the seasonal variations in wind speed over coastal areas of Høvsøre, Denmark, which shows that the wind speed variation is primarily dominant in the spring and winter season [43]. Also, it is found that the wind power ramp events are characterized based on magnitude error, phase error, and location error as pointed out by Potter et al. [44]. Cutler et al. have discussed the MesoLAPS and Wind power prediction tool (WPPT) for analyzing and categorizing large wind power ramp events. The root mean squared error (RMSE) is evaluated over a period of one year for two wind speed time series with 5-min and 10-min sampling intervals [45].

Results reveal that the RMSE values are found to be more optimistic than conventional persistence and climatology methods. Cornejo-Bueno et al. [46] have described the machine learning techniques like support vector regression (SVR), Gaussian process regression (GPR), multilayer perceptrons (MLPs), and extreme learning machines (ELMs) to forecast wind power ramp events. Experimental analysis is carried out for three wind farms, and results reveal that GPR-based regression yields better ramp forecasts for a time interval of 6 h.

7.3.1 Case study for onshore and offshore wind farms

In this section, we discuss the forecasting performance of the prediction models (persistence, ε-SVR, LSSVR, TSVR, ε-TSVR, RFR, and GBM). The ramp-up and ramp-down events are identified for a threshold of 10% of nominal wind power. The forecasting process is carried out by splitting the entire dataset into training (80%) and testing (20%) sets. For the persistence model, the forecast is carried out using two previous dispatch windows. Further, for SVR and its variants, the hyperparameters, that is, RBF bandwidth σ and regularization parameter C are tuned from the set $[2^{-10}, 2^{-9}, ..., 2^9, 2^{10}]$. For random forest regression (RFR) and gradient boosted machines (GBM), the simulations are carried out in R studio using random forest package. The number of trees used in training phase are 1000 for RFR, whereas for GBM, the learning rate is kept at 0.05, and the number of trees is 10000. The input feature set to all the prediction models is a matrix consisting of approximation signal A5 and detail signals (D1, D2, ..., D5) obtained from wavelet decomposition of wind speed time series. Further, we evaluate the absolute error for ramp-up and ramp-down events as R^{up} and R^{down}, respectively. Table 7.3 depicts the mean and standard deviation (SD) for the onshore and offshore wind speed datasets.

Table 7.3 Description of wind farm datasets for the month March 2019.

Onshore wind farm	Site coordinates	Mean	SD
Amakhala Emoyeni, SA (**a**)	−32.17° N, 25.95° E	6.264	3.198
Clyde, Scotland (**b**)	55.46° N, 03.65° W	3.829	1.626
Gansu, China (**c**)	40.20° N, 96.90° E	4.000	2.477
McCain Foods, UK (**d**)	52.56° N, 0.172° W	6.491	3.519
Shephards Flat, USA (**e**)	45.70° N, 120.06° W	6.074	2.618
Akhfenir, Morocco (**f**)	27.95° N, 12.00° W	3.096	1.505
Offshore wind farm	**Site coordinates**	**Mean**	**SD**
Amrumbank, Germany (**Aa**)	54.50° N, 7.80° E	11.176	4.962
Anholt, Denmark (**Bb**)	56.60° N, 11.21° E	8.999	3.537
Gemini, Netherlands (**Cc**)	54.03° N, 5.96° E	7.577	4.174
HornsRev 2, Denmark (**Dd**)	55.60° N, 7.59° E	11.183	4.546
Veja Mate, Germany (**Ee**)	54.31° N, 5.87° E	11.490	4.685
Walney, UK (**Ff**)	54.04° N, 5.92° W	11.342	5.015

For onshore wind farm sites, the datasets labeled are **a**, **b**, **c**, **d**, **e**, and **f**. For dataset **a**, GBM outperforms all the prediction models in terms of RMSE and NMSE. For predicting ramp-up and ramp-down events, TSVR outperforms GBM and RFR. In terms of R^2, TSVR gives the best fit close to 1.00, whereas ε-SVR gives a value greater than 1.00 causing over-fitting. The over-fitting in prediction is avoided by using RFR and GBM. Figs. 7.10–7.12 illustrate scatter plots for the different onshore datasets.

Similarly, for offshore wind farm datasets (**Aa**, **Bb**, **Cc**, **Dd**, **Ee** & **Ff**), Figs. 7.13–7.15 illustrate scatter plots.

TSVR outperforms all the models in terms of RMSE and NMSE for predicting wind speed, whereas for predicting ramp events, TSVR gives the best result in terms of R^2. Further, since the datasets are large sized, TSVR, ε-TSVR, RFR, and GBM take less computation time than conventional ε-SVR. The detailed performance metrics for onshore wind farm sites are depicted in Tables 7.4, 7.5, and 7.6.

Among different regressors, TSVR gives a minimum prediction error for ramp-up and ramp-down events for datasets **c** and **d**. In terms of RMSE, for dataset **c**, TSVR outperforms ε-TSVR, LSSVR, RFR, and GBM by 85%, 84.2%, 98.3%, and 97.6%, respectively.

From the performance metrics of the onshore wind sites we can observe that the decision tree ensemble methods like random forest and gradient boosted machines result in significantly lower RMSE and NMSE than supervised machine learning models. For example, for dataset **a**, ε-SVR

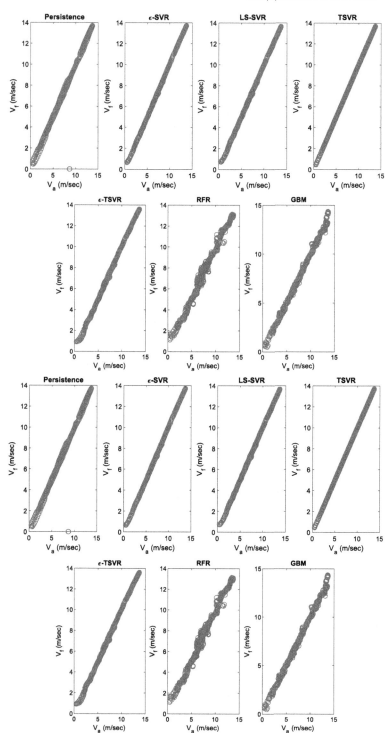

Figure 7.10 Scatter plots for onshore wind farm sites: Datasets **a** and **b**.

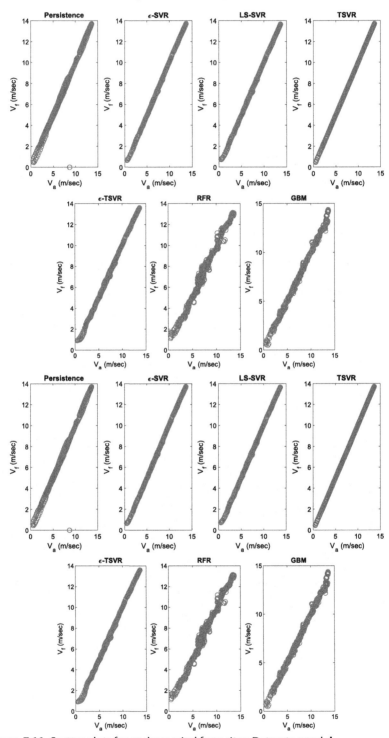

Figure 7.11 Scatter plots for onshore wind farm sites: Datasets **c** and **d**.

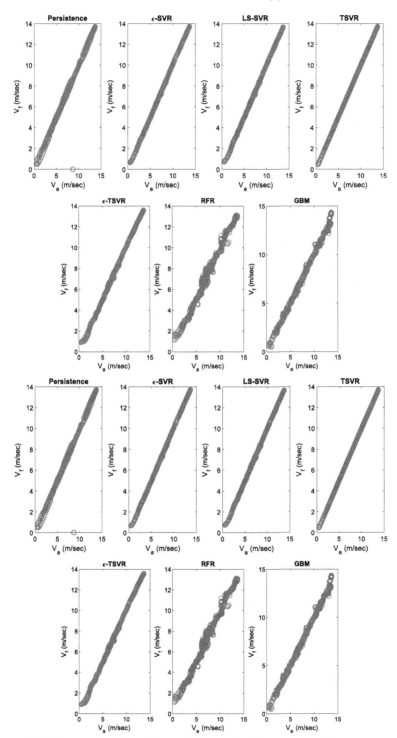

Figure 7.12 Scatter plots for onshore wind farm sites: Datasets **e** and **f**.

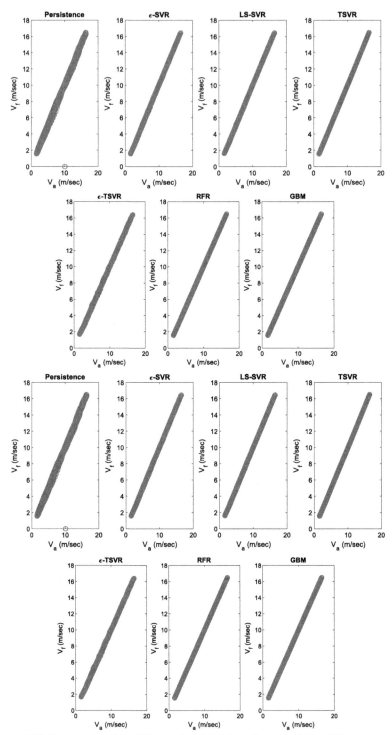

Figure 7.13 Scatter plots for offshore wind farm sites: Datasets **Aa** and **Bb**.

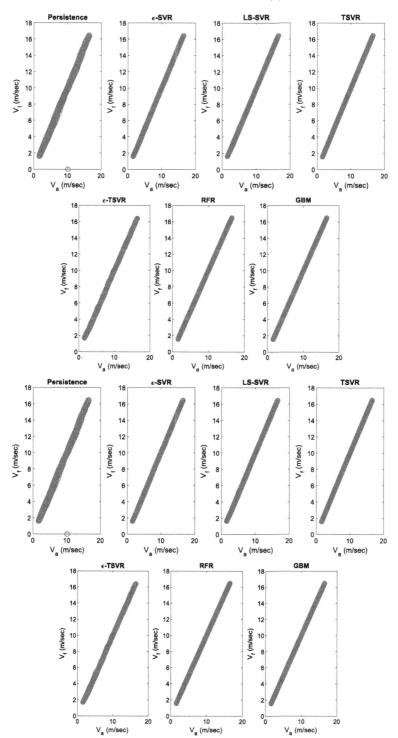

Figure 7.14 Scatter plots for offshore wind farm sites: Datasets **Cc** and **Dd**.

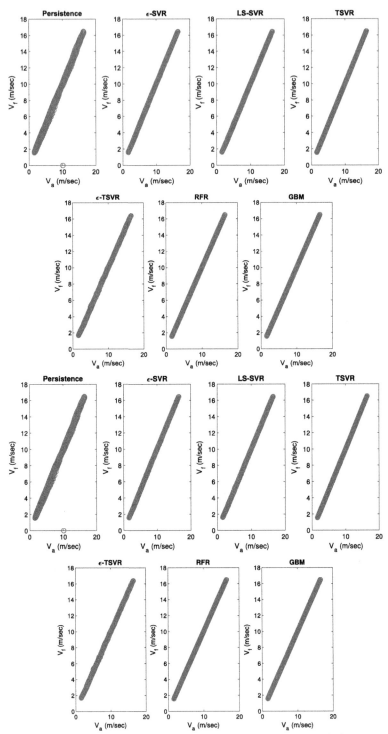

Figure 7.15 Scatter plots for offshore wind farm sites: Datasets **Ee** and **Ff**.

Table 7.4 Performance metrics of ramp prediction models (Onshore sites).

Dataset	Model	RMSE U2	NMSE R^{up}	R^2 R^{down}	U1 CPU time
a	LSSVR	0.3602	0.0177	0.9856	0.0024
		0.5678	0.0236	NA	1.46082
	TSVR	0.3525	0.0177	0.9899	0.0001
		0.0519	0.0016	NA	57.41689
	ε-TSVR	0.3475	0.0178	0.9701	0.0046
		0.6538	0.0036	NA	7.3946
	RFR	0.3436	0.1251	0.9506	0.5633
		0.6276	0.0789	NA	4.2693
	GBM	0.1197	0.0044	0.9705	0.0115
		0.1539	0.0043	NA	0.7119
	Persistence	0.4399	0.0221	1.0145	0.0259
		0.6296	0.2800	NA	0.02545
	ε-SVR	0.3620	0.0178	1.0004	0.0029
		6.6035	0.0862	NA	238.8274
b	LSSVR	0.0403	0.00057	0.9748	0.0044
		0.5716	0.1169	0.1498	0.76059
	TSVR	0.0268	0.0002	0.9832	0.0029
		0.444	0.0881	0.1228	23.9192
	ε-TSVR	0.0486	0.0008	0.9758	0.0053
		0.6978	0.1098	0.1332	3.3806
	RFR	0.06991	0.00121	0.9730	0.0660
		0.7191	0.1112	0.1421	32.11
	GBM	0.0651	0.0017	0.9986	0.0078
		0.0219	0.0235	0.0307	45.93
	Persistence	0.4399	0.0108	1.0068	0.0190
		0.6429	0.2350	0.2000	0.04356
	ε-SVR	0.0593	0.0013	0.9556	0.0065
		0.6517	0.1876	0.2367	84.4812

model gives an R^2 value greater than 1, which suggests overfitting when performed on an unseen data. The problem of overfitting as discussed in Chapter 5 can be eliminated by using a boosting algorithm like *XBoost*. Also it is worth noting that the error in predicting ramp-down event is greater than that in predicting ramp-up event. This observation remains true for most of the datasets.

Consider dataset **Aa**. The RMSE value is minimum for ε-TSVR followed by TSVR and LSSVR. The decision tree ensemble methods give a

Table 7.5 Performance metrics of ramp prediction models (Onshore sites).

Dataset	Model	RMSE	NMSE	R^2	U1
		U2	R^{up}	R^{down}	CPU time
c	LSSVR	0.1806	0.0026	0.9263	0.0141
		0.2834	0.0600	0.5072	1.41105
	TSVR	0.0284	0.00002	0.9884	0.0022
		0.0034	0.0014	0.1119	61.54305
	ε-TSVR	0.1964	0.0031	0.9252	0.0154
		0.5499	0.0793	0.5588	7.82425
	RFR	1.7407	0.2461	0.4191	0.1478
		1.3082	1.7112	5.0778	8.3472
	GBM	1.2148	0.1239	0.6060	0.1740
		0.2787	0.1661	3.4109	0.6997
	Persistence	0.3989	0.0129	1.0051	0.0309
		0.6207	0.3250	0.1950	0.03783
	ε-SVR	0.1937	0.0030	0.9253	0.0152
		0.4438	0.0563	0.5227	251.86508
d	LSSVR	0.0263	0.000009	0.9927	0.0024
		0.4229	0.0300	0.0109	0.519493
	TSVR	0.0040	0.00002	1.000	0.0003
		0.0662	0.0004	0.00002	15.967091
	ε-TSVR	0.0732	0.0007	0.9814	0.0067
		0.6507	0.1065	0.0471	2.067738
	RFR	0.6014	0.0489	0.7966	0.0543
		0.9854	0.0204	0.1416	5.187355
	GBM	0.2758	0.0103	0.9585	0.0252
		0.1458	0.1001	0.1444	0.505211
	Persistence	0.2154	0.0063	1.0023	0.0197
		0.5417	0.3550	0.3500	0.0616
	ε-SVR	0.0382	0.0001	0.9951	0.0035
		0.6014	0.0594	0.0345	33.06520

good predictive performance when compared to the persistence algorithm. In terms of R^2, the persistence and ε-SVR models give a value greater than 1, which causes the data to overfit. Using decision tree methods, the overfitting can be reduced as seen by the R^2 values of RFR and GBM models. Similar observation can be found for dataset **Bb** as well.

It is worthwhile to note that ramp prediction performance for GBM is significantly closer to TSVR in case of offshore wind farms as indicated in Tables 7.7, 7.8, and 7.9.

Table 7.6 Performance metrics of ramp prediction models (Onshore sites).

Dataset	Model	RMSE / U2	NMSE / R^{up}	R^2 / R^{down}	U1 / CPU time
e	LSSVR	0.0232	0.000006	0.9935	0.0017
		0.3551	0.0626	0.0689	1.022718
	TSVR	0.0042	0.00002	0.9994	0.00004
		0.004	0.0134	0.0125	43.252339
	ε-TSVR	0.0344	0.0001	0.9921	0.0025
		0.7606	0.1058	0.1035	5.18525
	RFR	0.09609	0.00137	0.9607	0.0016
		0.2612	0.1527	0.093129	15.87
	GBM	0.07739	0.0008	0.9968	0.0059
		1.3718	0.0822	0.0160	36.72
	Persistence	0.2517	0.0077	1.0065	0.0185
		0.6629	0.5450	0.1700	0.002612
	ε-SVR	0.0400	0.0001	0.9974	0.0029
		0.4500	0.0603	0.0981	253.34162
f	LSSVR	0.0216	0.0001	0.9971	0.0022
		0.4651	0.0475	0.1055	0.700270
	TSVR	0.0030	0.000002	0.9993	0.0003
		0.0112	0.0076	0.0003	17.6919
	ε-TSVR	0.0467	0.0005	0.9922	0.0049
		0.8286	0.1660	0.0980	4.27225
	RFR	0.085906	0.00175	0.9552	0.0057
		0.8316	0.03032	0.08309	15.83
	GBM	0.07239	0.001219	0.9998	0.0051
		0.7216	0.01447	0.1298	44.66
	Persistence	0.2250	0.0118	1.0040	0.0234
		0.5833	0.3050	0.2250	0.0494
	ε-SVR	0.0440	0.0004	1.0129	0.0046
		0.7638	0.0518	0.0776	63.81193

7.3.2 Discussion on uncertainties in ramp events

Results obtained so far depict that SVR-based regressors have an upper hand over RFR and GBM based prediction methods. It is worth noting that in the study ramp events the time interval between consecutive samples plays an important role in deciding the threshold value. In a study carried out by Ouyang et al., the wind speed data collected is sampled every 15 minutes for a wind farm site in China, and ARIMA model is used to forecast wind power [47]. Further, the swinging door algorithm is used

Table 7.7 Performance metrics of ramp prediction models (Offshore sites).

| Dataset | Model | RMSE | NMSE | R^2 | U1 |
		U2	R^{up}	R^{down}	CPU time
Aa	LSSVR	0.0225	0.00004	0.9899	0.0013
		0.352	0.004	0.0221	0.507431
	TSVR	0.0037	0.00001	0.9998	0.0002
		0.0745	0.0034	0.0038	11.28299
	ε-TSVR	0.0429	0.0001	0.9891	0.0025
		0.5738	0.0162	0.0777	2.0435
	RFR	0.0989	0.00104	0.9799	0.0036
		0.5161	0.1032	0.05638	31.2
	GBM	0.0967	0.0075	0.9904	0.0032
		0.5016	0.1052	0.0423	50.59
	Persistence	0.324	0.0084	1.0001	0.019
		0.6	0.185	0.345	0.051183
	ε-SVR	0.0367	0.0001	1.0009	0.0022
		0.5024	0.0167	0.0375	84.006
Bb	LSSVR	0.0198	0.00006	0.9912	0.0013
		0.3908	0.008	0.031	0.4999
	TSVR	0.0038	0.000002	0.9998	0.0002
		0.0115	0.0043	0.0028	12.5415
	ε-TSVR	0.0434	0.0003	0.9998	0.0028
		0.7492	0.1064	0.0771	1.9274
	RFR	0.7733	0.00105	0.9718	0.0034
		0.0081	0.18005	0.07037	31.0600
	GBM	0.06248	0.00068	0.9994	0.003
		0.0071	0.180035	0.07988	44.5800
	Persistence	0.1655	0.0048	1.0067	0.0106
		0.6389	0.38	0.3	0.1496
	ε-SVR	0.0367	0.0002	1.0001	0.0024
		0.0074	0.0186	0.0849	27.9990

to detect ramp events. The RMSE values for the ramp event prediction are found to be in the range of 32%. Further, using machine learning techniques such as classical SVR, GPR, MLP, and ELM, the ramp events are predicted for a time interval 6 h, and RMSE values are found to be in the range 5–7 MW).

However, wind speed time series of onshore and offshore wind farms has a lot of variability in terms of magnitude. Thus a ramp event study with 10 minutes as time interval poses critical impositions in market clearing and

Table 7.8 Performance metrics of ramp prediction models (Offshore sites).

Dataset	Model	RMSE U2	NMSE R^{up}	R^2 R^{down}	U1 CPU time
Cc	LSSVR	0.0232	0.00006	0.9944	0.0021
		0.3838	0.0217	0.0305	0.514049
	TSVR	0.0064	0.000004	0.9993	0.0005
		1.0497	0.0001	0.0012	13.034972
	ε-TSVR	0.0463	0.0002	0.9871	0.0043
		1.0083	0.056	0.076	1.90010
	RFR	0.07823	0.00724	0.9763	0.0051
		1.0091	0.7805	0.0896	30.58
	GBM	0.07146	0.0039	0.9863	0.0041
		1.0012	0.1687	0.04527	43.47
	Persistence	0.2577	0.0079	1.0015	0.0238
		0.5	0.55	0.555	0.04261
	ε-SVR	0.0626	0.0004	0.9875	0.0058
		1.0347	0.0765	0.0866	39.806
Dd	LSSVR	0.0244	0.000004	0.9987	0.0014
		0.057	0.0139	0.0119	0.5023
	TSVR	0.0032	0.00007	0.9999	0.0001
		0.0213	0.0024	0.0027	12.7748
	ε-TSVR	0.0352	0.00009	0.9994	0.002
		1.5215	0.0338	0.0229	1.9140
	RFR	0.09762	0.00072	0.9758	0.0042
		0.7166	0.09793	0.04394	38.7200
	GBM	0.0793	0.001004	0.9868	0.0056
		0.8166	0.00686	0.0029315	55.4600
	Persistence	0.4252	0.0138	1.0038	0.0236
		0.75	0.055	0.45	0.0492
	ε-SVR	0.0379	0.0001	1.0105	0.0021
		0.6116	0.0136	0.0123	80.1100

day-ahead scenarios. The current work deals with predicting wind power ramp events with 10-min sampling interval for onshore and offshore wind sites. With offshore wind farms, the wind speed being high and variable, the probability of ramp event increases. Thus, in the current work, we extend the ramp event prediction study by incorporating advanced machine learning algorithms like variants of SVR, random forest regression, and gradient boosted machines. Results reveal that the TSVR-based prediction model yields the lowest error for ramp-up and ramp-down events.

Table 7.9 Performance metrics of ramp prediction models (Offshore sites).

Dataset	Model	RMSE U2	NMSE R^{up}	R^2 R^{down}	U1 CPU time
Ee	LSSVR	0.0263	0.00005	0.9994	0.0015
		0.4724	0.0032	0.0626	0.50647
	TSVR	0.004	0.00001	0.9997	0.0002
		0.0309	0.0004	0.0018	12.743
	ε-TSVR	0.0437	0.0001	0.9994	0.0024
		0.633	0.004	0.1313	1.90508
	RFR	0.1059	0.0008	0.9761	0.0059
		0.8372	0.2281	0.3498	37.43
	GBM	0.08278	0.0004	0.9951	0.0096
		1.8164	0.08923	0.0112	57.2
	Persistence	0.3866	0.0109	1.0057	0.0215
		0.6	0.19	0.105	0.052525
	ε-SVR	0.0338	0.00008	1.0041	0.0019
		0.1171	0.0885	0.0112	90.5178
Ff	LSSVR	0.0206	0.00003	0.9991	0.0012
		0.3587	0.0015	0.044	0.58391
	TSVR	0.0029	0.000006	0.9995	0.0001
		0.173	0.0003	0.0065	12.78991
	ε-TSVR	0.037	0.0001	0.9912	0.0022
		0.419	0.0034	0.0408	1.87901
	RFR	0.1059	0.00084	0.9774	0.0062
		1.2043	0.1264	0.2182	33.01
	GBM	0.0929	0.00064	0.9976	0.0054
		0.3996	0.04356	0.07129	48.38
	Persistence	0.3244	0.0079	1.0036	0.0139
		0.5	0.21	0.405	0.05162
	ε-SVR	0.033	0.00008	1.0001	0.0019
		35.36	0.0358	0.0535	87.76891

A ramp event in onshore and offshore wind farms is a challenging issue. The randomness in wind speed stimulated by turbulent air flow can cause unwanted vibrations in turbine blade and tower, thus questioning its structural stability. Randomness in a signal or time series can often be expressed by calculating the log energy entropy. To critically examine a ramp event signal expressed as (7.24), we decompose the obtained signal by two signal processing techniques such as wavelet transform decomposition (WT) and empirical model decomposition (EMD). Whereas the wavelet transform

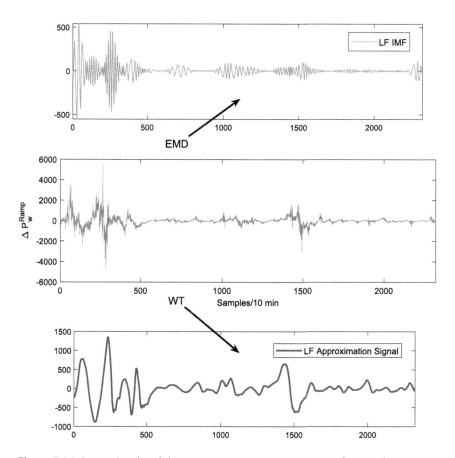

Figure 7.16 Ramp signal and decompositions using wavelet transform and EMD.

decomposes a signal into low-frequency and high-frequency components, EMD, on the other hand, is capable of extracting oscillatory features from the signal for a dominant frequency. The ramp signal is decomposed, using db4 wavelet filter and wavelet transform, into five intrinsic mode functions (IMFs) using EMD. The log energy entropy for a signal $h(t)$ is given as

$$E\{h(t)\} = \sum_{t=0}^{T} \log(h(t))^2. \tag{7.25}$$

A typical ramp event signal and its signal decomposition using the wavelet transform and EMD are illustrated in Fig. 7.16, which shows the low-frequency (LF) signals obtained from the wavelet transform and EMD.

Table 7.10 Log energy entropy for ramp event signals.

Dataset	Model	Log energy entropy	
		WT	EMD
a	TSVR	6.6094×10^3	5.6985×10^3
	RFR	6.2866×10^3	5.5948×10^3
	GBM	6.7926×10^3	5.9943×10^3
Aa	TSVR	1.7259×10^4	1.5646×10^4
	RFR	2.0676×10^4	1.2297×10^4
	GBM	2.0776×10^4	1.4621×10^4

Such signal decompositions are used to assess the randomness in the ramp event signal by calculating the log energy entropy of predicted ramp signals based on TSVR, RFR, and GBM as shown in Table 7.10.

For a particular prediction model, the log energy entropy based on WT is greater than that of EMD, suggesting higher-order randomness in the ramp event signal. As far as uncertainties are concerned, the decomposition-based prediction models yield better ramp prediction than single methods as listed in the literature. In the current work, wind power ramp events are studied for six onshore and six offshore wind farm sites. Wind speed data for March 2019 with 10-min sampling interval are collected. The available data are transformed to a hub height of 90 m for all the datasets. The threshold value is chosen as 10% of nominal wind power, based on which the ramp event points are assessed for absolute error. The machine learning-based prediction models are compared with the benchmark persistence model. TSVR-based model gives minimum error for both ramp-up and ramp-down events. The coefficient of performance (R^2) for RFR and GBM in case of offshore wind farm datasets is found to be close to 1.00, which indicates a good agreement between predicted and actual values. Further, it is observed that the error in predicting ramp-down event is greater than in ramp-up event.

Overall, machine learning-based prediction models are good approximators for analyzing short-term wind ramps. Randomness in wind power ramp series in case of TSVR, RFR, and GBM is evaluated using signal decomposition techniques, and with RFR the randomness is minimal. Machine intelligent models like TSVR and GBM affirm a stable grid operation in the presence of large ramp-up and ramp-down events. Among the regressors, ε-TSVR outperforms TSVR, LS-SVR, and ε-SVR in terms of absolute error. Ramp events are analyzed for different hub heights, and the number of recorded ramp events increased significantly with height. Machine intelligent hybrid methodology improves the forecasting performance of wind farms during ramp events.

References

[1] P. Breeze, Wind Power Generation, Academic Press, 2015.

[2] J. Waewsak, C. Kongruang, Y. Gagnon, Assessment of wind power plants with limited wind resources in developing countries: application to Ko Yai in southern Thailand, Sustainable Energy Technologies and Assessments 19 (2017) 79–93, https://doi.org/10.1016/j.seta.2016.12.001.

[3] S. Ji, B. Chen, Carbon footprint accounting of a typical wind farm in China, Applied Energy 180 (2016) 416–423, https://doi.org/10.1016/j.apenergy.2016.07.114.

[4] H.-J. Wagner, J. Mathur, Introduction to Wind Energy Systems, Springer Berlin Heidelberg, 2009.

[5] K. Dai, A. Bergot, C. Liang, W.-N. Xiang, Z. Huang, Environmental issues associated with wind energy – a review, Renewable Energy 75 (2015) 911–921, https://doi.org/10.1016/j.renene.2014.10.074.

[6] M. Desholm, Avian sensitivity to mortality: prioritising migratory bird species for assessment at proposed wind farms, Journal of Environmental Management 90 (8) (2009) 2672–2679, https://doi.org/10.1016/j.jenvman.2009.02.005.

[7] D. Hayashi, J. Huenteler, J.I. Lewis, Gone with the wind: a learning curve analysis of China's wind power industry, Energy Policy 120 (2018) 38–51, https://doi.org/10.1016/j.enpol.2018.05.012.

[8] V. Pérez-Andreu, C. Aparicio-Fernández, A. Martínez-Ibernón, J.-L. Vivancos, Impact of climate change on heating and cooling energy demand in a residential building in a Mediterranean climate, Energy 165 (2018) 63–74, https://doi.org/10.1016/j.energy.2018.09.015.

[9] M.A. Martins, J. Tomasella, D.A. Rodriguez, R.C. Alvalá, A. Giarolla, L.L. Garofolo, J.L.S. Júnior, L.T. Paolicchi, G.L. Pinto, Improving drought management in the Brazilian semiarid through crop forecasting, Agricultural Systems 160 (2018) 21–30, https://doi.org/10.1016/j.agsy.2017.11.002.

[10] P. Calanca, Weather forecasting applications in agriculture, in: Encyclopedia of Agriculture and Food Systems, Elsevier, 2014, pp. 437–449.

[11] M. Eberhardt, D. Vollrath, The effect of agricultural technology on the speed of development, World Development 109 (2018) 483–496, https://doi.org/10.1016/j.worlddev.2016.03.017.

[12] A.G. Maia, B.C.B. Miyamoto, J.R. Garcia, Climate change and agriculture: do environmental preservation and ecosystem services matter? Ecological Economics 152 (2018) 27–39, https://doi.org/10.1016/j.ecolecon.2018.05.013.

[13] A. Ullrich, M. Volk, Application of the soil and water assessment tool (SWAT) to predict the impact of alternative management practices on water quality and quantity, Agricultural Water Management 96 (8) (2009) 1207–1217, https://doi.org/10.1016/j.agwat.2009.03.010.

[14] S.K. Dubey, D. Sharma, Assessment of climate change impact on yield of major crops in the Banas river basin, India, Science of the Total Environment 635 (2018) 10–19, https://doi.org/10.1016/j.scitotenv.2018.03.343.

[15] P. Robichaud, J. Jennewein, B. Sharratt, S. Lewis, R. Brown, Evaluating the effectiveness of agricultural mulches for reducing post-wildfire wind erosion, Aeolian Research 27 (2017) 13–21, https://doi.org/10.1016/j.aeolia.2017.05.001.

[16] H. Li, J. Tatarko, M. Kucharski, Z. Dong, PM2.5 and PM10 emissions from agricultural soils by wind erosion, Aeolian Research 19 (2015) 171–182, https://doi.org/10.1016/j.aeolia.2015.02.003.

[17] E. Sirjani, A. Sameni, A.A. Moosavi, M. Mahmoodabadi, B. Laurent, Portable wind tunnel experiments to study soil erosion by wind and its link to soil properties in the Fars province, Iran, Geoderma 333 (2019) 69–80, https://doi.org/10.1016/j.geoderma.2018.07.012.

[18] G. Wiggs, P. Holmes, Dynamic controls on wind erosion and dust generation on West-central free state agricultural land, South Africa, Earth Surface Processes and Landforms 36 (6) (2010) 827–838, https://doi.org/10.1002/esp.2110.

[19] F. Senturk, Hydraulics of Dams and Reservoirs, Water Resources Pubns, 1995.

[20] J. Wang, W. Yang, P. Du, Y. Li, Research and application of a hybrid forecasting framework based on multi-objective optimization for electrical power system, Energy 148 (2018) 59–78.

[21] B. Pushpawela, R. Jayaratne, L. Morawska, The influence of wind speed on new particle formation events in an urban environment, Atmospheric Research 215 (2019) 37–41, https://doi.org/10.1016/j.atmosres.2018.08.023.

[22] J. Chen, H. Shi, B. Sivakumar, M.R. Peart, Population, water, food, energy and dams, Renewable & Sustainable Energy Reviews 56 (2016) 18–28, https://doi.org/10.1016/j.rser.2015.11.043.

[23] A. Peña, S.-E. Gryning, C.B. Hasager, Measurements and modelling of the wind speed profile in the marine atmospheric boundary layer, Boundary-Layer Meteorology 129 (3) (2008) 479–495, https://doi.org/10.1007/s10546-008-9323-9.

[24] R.J. Keenan, G.A. Reams, F. Achard, J.V. de Freitas, A. Grainger, E. Lindquist, Dynamics of global forest area: results from the FAO global forest resources assessment 2015, Forest Ecology and Management 352 (2015) 9–20, https://doi.org/10.1016/j.foreco.2015.06.014.

[25] Z. Jiao-jun, L. Xiu-fen, G. Yutaka, M. Takeshi, Wind profiles in and over trees, Journal of Forestry Research 15 (4) (2004) 305–312, https://doi.org/10.1007/bf02844959.

[26] P. Hannah, J.P. Palutikof, C.P. Quine, Predicting windspeeds for forest areas in complex terrain, in: M.P. Coutts, J. Grace (Eds.), Wind and Trees, Cambridge University Press, 1995, pp. 113–130.

[27] J.J. Finnigan, Y. Brunet, Turbulent airflow in forests on flat and hilly terrain, in: M.P. Coutts, J. Grace (Eds.), Wind and Trees, Cambridge University Press, 1995, pp. 3–40.

[28] R. Simpson, A model to control emissions which avoid violations of PM10 health standards for both short and long term exposures, Atmospheric Environment. Part A, General Topics 24 (4) (1990) 917–924, https://doi.org/10.1016/0960-1686(90)90294-w.

[29] Airnow department of state, https://www.airnow.gov, 2018.

[30] B. Greaves, J. Collins, J. Parkes, A. Tindal, Temporal forecast uncertainty for ramp events, Wind Engineering 33 (4) (2009) 309–319.

[31] C. Kamath, Understanding wind ramp events through analysis of historical data, in: IEEE PES T&D 2010, 2010, pp. 1–6.

[32] G. Ren, J. Liu, J. Wan, Y. Guo, D. Yu, Overview of wind power intermittency: impacts, measurements, and mitigation solutions, Applied Energy 204 (2017) 47–65.

[33] Anemometer data (wind speed, direction) for Beresford, South Dakota (2006), https://openei.org, 2019. (Accessed 1 March 2019).

[34] J.S. Irwin, A theoretical variation of the wind profile power-law exponent as a function of surface roughness and stability, Atmospheric Environment (1967) 13 (1) (1979) 191–194.

[35] E.K. Akpinar, S. Akpinar, An assessment of wind turbine characteristics and wind energy characteristics for electricity production, Energy Sources, Part A: Recovery, Utilization, and Environmental Effects 28 (10) (2006) 941–953.

[36] J. Cruz, M. Atcheson (Eds.), Floating Offshore Wind Energy, Springer International Publishing, 2016.

[37] M.D. Esteban, J.J. Diez, J.S. López, V. Negro, Why offshore wind energy? Renewable Energy 36 (2) (2011) 444–450, https://doi.org/10.1016/j.renene.2010.07.009.

[38] Aerodynamic roughness length –AMS glossary, http://glossary.ametsoc.org, 2019. (Accessed 28 April 2019).

[39] S. Colwell, B. Basu, Tuned liquid column dampers in offshore wind turbines for structural control, Engineering Structures 31 (2) (2009) 358–368, https://doi.org/10.1016/j.engstruct.2008.09.001.

[40] C.L. Vincent, P. Pinson, G. Giebela, Wind fluctuations over the North Sea, International Journal of Climatology 31 (11) (2011) 1584–1595, https://doi.org/10.1002/joc.2175.

[41] J. Gjerstad, S.E. Aasen, H.I. Andersson, I. Brevik, J. Løvseth, An analysis of low-frequency maritime atmospheric turbulence, Journal of the Atmospheric Sciences 52 (15) (1995) 2663–2669.

[42] O. Kramer, N.A. Treiber, M. Sonnenschein, Wind Power Ramp Event Prediction With Support Vector Machines, Lecture Notes in Computer Science, Springer International Publishing, 2014, pp. 37–48.

[43] J. Nissen, On the Application of a Numerical Model to Simulate the Coastal Boundary Layer, Ph.D. thesis, 2008.

[44] C.W. Potter, E. Grimit, B. Nijssen, Potential benefits of a dedicated probabilistic rapid ramp event forecast tool, in: 2009 IEEE/PES Power Systems Conference and Exposition, IEEE, 2009.

[45] N. Cutler, M. Kay, K. Jacka, T.S. Nielsen, Detecting, categorizing and forecasting large ramps in wind farm power output using meteorological observations and WPPT, Wind Energy 10 (5) (2007) 453–470, https://doi.org/10.1002/we.235.

[46] L. Cornejo-Bueno, L. Cuadra, S. Jiménez-Fernández, J. Acevedo-Rodríguez, L. Prieto, S. Salcedo-Sanz, Wind power ramp events prediction with hybrid machine learning regression techniques and reanalysis data, Energies 10 (11) (2017) 1784, https://doi.org/10.3390/en10111784.

[47] T. Ouyang, X. Zha, L. Qin, Y. He, Z. Tang, Prediction of wind power ramp events based on residual correction, Renewable Energy 136 (2019) 781–792, https://doi.org/10.1016/j.renene.2019.01.049.

CHAPTER 8

Supervised learning for forecasting in presence of wind wakes

Wind energy installations require a precise study of land area available and nearby atmospheric conditions. Power captured from wind resource is dependent on wind speed at the site. Optimal placement of wind turbines in a wind farm for maximum power capture in the presence of wind wakes is a major challenge. Wind wakes represent an aerodynamic phenomenon causing reduced power captured at downstream turbine. In this chapter we formulate a bilateral Gaussian wake model-based approach using Jensen's and Frandsen's Gaussian variations of wind wake models. The proposed model ensures that the wake effect of the incident wind speed on a downwind turbine is minimized as compared to existing benchmark analytical models for single and multiple wake scenarios. Furthermore, short-term wind speed forecasting in the presence of wakes is carried out for two wind farm layouts considering benchmark wake models and our proposed model. The significant upwind turbines are identified using a method called Grey relational analysis, and the forecasting accuracy is evaluated.

8.1 Introduction

Growing energy demands are rapidly facilitating the wind turbine installations globally in the form of large wind parks. Wind turbines are installed in large land mass to convert the energy available from moving air to electrical energy. Due to constrained land area and cost of the equipment, we have to design a proper wind farm layout for the required energy generation [1], [2]. Wind power capture by wind turbine is affected by many factors like wind speed, wind direction, and optimal turbine spacing [3].

In a wind farm terrain, wind turbines must be placed at an optimal operating distance from each other to avoid potential derating caused by wind wakes. Wind wake is an aerodynamic phenomenon leading to (a) reduction in wind speed magnitude at the downwind turbine and (b) increased air turbulence causing mechanical loading on the turbine structure [4]. Wind wakes can be classified as near-end wakes and far-end wakes; the former

Supervised Machine Learning in Wind Forecasting and Ramp Event Prediction
https://doi.org/10.1016/B978-0-12-821353-7.00019-3

extends up to a distance of one to three rotor diameters where the flow is dependent on the turbine geometry [5].

Early wake models evolved in the 1980s with N.O. Jensen proposing a single wake model describing wind speed deficit caused by a single up-wind turbine on downwind turbine. Various analytical models like Ainslie's model [6], Larsen's model [7], and Frandsen's model [8] have been used. The most commonly used model is Jensen's model, which assumes linear wake expansion after the hub. Ainslie's model considers a numerical wake model with symmetric Reynolds equation to compute wake development such that the wake deficit decays monotonically with increasing down-stream distance (experimentally 4D) [9]. However, the computation time of Ainslie's model is found to be large [10]. A new 2D Jensen wake model is proposed by Tian et al., which incorporates the variable wake decay rate rather than a constant one. Numerical simulations are performed for computing the wake deficit and are compared with field measurements. Results reveal that the proposed 2D Jensen wake model underestimates for near-wake regions [11]. Ishihara et al. [12] have presented an analytical model that encapsulates the effect of thrust coefficient and air turbulence on the wake deficit. The numerical simulations are compared with a test carried out in a wind tunnel, and the results of the proposed analytical model are in good agreement with experimental analysis.

Wind wakes causing power loss for an individual wind turbine lead to a detailed study of wind wakes, and hence over the years many analytical and field models have been developed to study them [13], [14]. Experimental results have shown that due to wake interference, the downwind turbine experiences up to 40% of power loss and 80% of increased dynamic loading on the turbine structure [15]. Here an experimental setup for wind turbine with rotor diameter 0.9 m is placed 4 rotor diameters from the wind inlet section, and the wake velocity distribution is measured at a downstream distance of 0.6D and 3D. Jensen's wake model is validated and tested for accommodating the power losses in wind, and the losses are found in acceptable range [16–18]. Wake study also plays an important role in the wind farm layout optimization problem (WFLOP), where optimal placement of the wind turbines leads to minimum wake effect and maximum power capture [19].

Wind being stochastic in its nature, its accurate forecasting is a major challenge in power industry. Wind speed forecasting has become an essential component to ensure power system security and reliability as increased wind power penetration leads to an unreliable operation [20]. It also serves the purpose of market clearing operations and efficient load dispatch planning. Wind forecasting is broadly categorized on the basis of prediction horizon, that is, very short-term (from a few seconds to 30 min), short-term (from 30 min to 6 h), medium term (from 6 h to 1 day), and long-term (from 1 day to 1 week) wind forecasting [21]. Okumus et al. [22] have reviewed recent forecasting schemes, which include hybrid methods for improving the accuracy of prediction. Among these, the most used forecasting methods are a combination of two or more machine learning methods combined with a time-series model (ARMA and ARIMA). Further, on the basis of methods of forecasting like time-series-based forecasting models like ARMA and ARIMA methods are employed for short-term wind speed and power forecasts. However, due to nonlinearity in wind speed series, machine learning approaches like ANNs, SVR [23], and ELM have been found in use recently. Patel et al. [2,24] have discussed forecasting algorithms for wind power dispatch for multiple wind farms considering battery life optimization.

Forecasting wind speed in the presence of wind wakes is an uphill task. Wind wakes cause reduction in power captured from wind resource, and thus optimal placement of wind turbines in a farm leads to efficient land area usage. In case of a wake affected downwind turbine the most significant upwind turbines affect the forecasting accuracy of the downwind turbine. In the present context, most significant upwind turbines are the turbines whose wake effect causes maximum power deficit at a downwind turbine. The major contributions of this chapter are as follows.

1. A bilateral Gaussian wake model comprising Jensen and Frandsen components is proposed and tested for single wake and multiple wake scenarios for two artificial wind farm layouts with two wind speed datasets.

2. Short-term wind speed forecasting is studied considering the wake effects, and significant upwind turbines are identified using Grey relational analysis. Forecasting accuracy is analyzed for proposed bilateral Gaussian wake model and benchmark models.

This chapter is organized as follows. Section 8.2 describes various wind wake models: Jensen's, Frandsen's, and the proposed bilateral Gaussian wake models. Further, Section 8.3 discusses the individual methodologies for short-term wind speed forecasting in the presence of wakes. In Section 8.4, we present results and discussions.

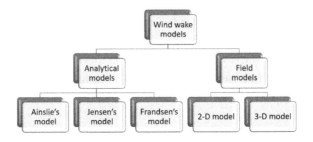

Figure 8.1 Categorization of different wake models.

8.2 Wind wakes

As already described, wind wakes lead to reduction in velocity at the down-wind turbine in a wind farm. Wake-affected wind turbine also suffers from the problem of increased mechanical loading on the turbine structure. Apart from analytical wake models cited in the literature like Jensen's model, Ainslie's model, Larsen's model, and Frandsen's model, many field models like 2D and 3D field models based on computational fluid dynamics (CFD) have also been put into wake modeling. The different wake models are categorized in Fig. 8.1.

Due to computational uncertainly and time consumption, analytical wake models are a preferred choice over field models [8]. We discuss analytical models for single wake and multiple wake conditions in a wind farm layout.

8.2.1 Jensen's and Frandsen's single wake models

Jensen [25] proposed a single wake model based on the assumption that the wake cone extends linearly with the downwind distance. The free-stream wind speed u_0 is the expected wind speed to be received by the downwind turbines, but due to wake effect, a velocity deficit is observed at these turbines. Based on the conservation of momentum of fluid across the wake cone, the radius of wake-affected wind turbine at a distance x from the upwind turbine can be expressed as

$$r_x = r_0 + \alpha x, \tag{8.1}$$

where r_0 is the rotor radius of the wind turbine, α represents the rate of wake expansion behind rotor, and x is the downwind distance as shown in Fig. 8.2.

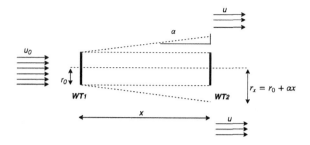

Figure 8.2 Wind turbine layout for Jensen's single wake model.

The choice of value of α depends on the wind farm terrain (surface roughness, hub height) and atmospheric conditions (air turbulence). An empirical estimation of wake expansion constant α is

$$\alpha = \kappa \left[\log_e \left(\frac{h}{z_0} \right) - \psi_m(h/L) \right]^{-1}, \qquad (8.2)$$

where $\kappa = 0.4$ is the von Karman constant, and h and z_0 are the hub height of turbine and surface roughness length. The factor $\psi_m(h/L)$ determines the local atmospheric stability correction at a given hub height [26]. The effective wind speed u_j at the downwind turbine located at a distance x due to wake effect of a single upwind turbine as per Jensen's model is given as

$$u_j = u_0 \left(1 + (\sqrt{1 - C_t} - 1) \left(\frac{r_0}{r_x} \right)^2 \right), \qquad (8.3)$$

where $C_t = a(2 - a)$ is the thrust coefficient with induction factor $a = 1 - \frac{u_j}{u_0}$ as per Actuator disk theory.

Frandsen et al. [8] have described the wake expansion immediately after rotor as rectangular profile compared to trapezoidal profile in Jensen's model. The wind speed u_f at a downwind distance x as per Frandsen's model is given as

$$u_f = u_0 \left(\frac{1}{2} \pm \frac{1}{2} \sqrt{1 - 2C_t \left(\frac{r_0}{r_x} \right)^2} \right). \qquad (8.4)$$

An empirical result was obtained to estimate the relationship between r_x and C_t, α, and r_0, which is given as

$$r_x = r_0 (\beta + \alpha x/2r_0)^{1/2}, \qquad (8.5)$$

$$\beta = 0.5 \left(\frac{1 + \sqrt{1 - C_t}}{\sqrt{1 - C_t}} \right). \tag{8.6}$$

In (8.4), the "+" sign is applicable for $a \leq 0.5$ and the "−" sign for $a > 0.5$.

8.2.2 Proposed model for wind wakes

The velocity deficit estimated by Jensen and Frandsen are based on the top-hat-like velocity distribution. Several wind tunnel experiments have shown that the velocity deficit in transverse direction for a far-wake region follows a Gaussian distribution [12]. Wu and Porté-Agel [27] have studied the turbulence effect on stand–alone turbine wakes based on large-eddy simulation (LES) framework. The velocity deficit at the far-wake end assumes a Gaussian distribution [28] such as

$$u = u_0 \left(1 - A(x) e^{\frac{-r^2}{2\sigma^2}} \right), \tag{8.7}$$

$$A(x) = 1 - \sqrt{1 - \frac{C_t}{8(\sigma/2r_0)^2}}, \tag{8.8}$$

where u_0 is the free-stream wind speed, $A(x)$ is the maximum normalized velocity deficit caused at each downstream position x, and σ is the wake width for each downwind distance x.

However, the mass flow rate between the wind turbines is not constant as certain percentage of incoming kinetic energy of wind is lost due to air turbulence. Another wake model proposed by Larsen was used to determine the speed of wind in wake affected downstream area, but due to high computational requirements, we do not prefer it here. Based on the Gaussian variation of Jensen's and Frandsen's wake models, we propose a bilateral Gaussian wake model shown in Fig. 8.3, where two wind turbines are placed apart "x" rotor diameter apart, and the velocity deficit assumes a Gaussian distribution at the far-end.

The wake expands linearly with downstream distance x and is dependent on atmospheric conditions in terms of stability factor measured by $\psi_m(h/L)$ and terrain conditions in terms of surface roughness length z_0. The boundary lines Y_1, Y_2, and Y_3 are the regions where the wake effect is analyzed based on Jensen's and Frandsen's original wake scenario. According to Jensen, in the boundaries Y_2 and Y_3 the law of conservation of mass

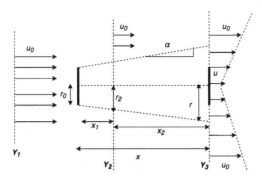

Figure 8.3 Bilateral Gaussian wake model.

holds, which can be mathematically expressed as

$$\int_{r_2}^{\infty} \rho u_0 2\pi \, r dr + \int_0^{r_2} \rho u_2 2\pi \, r dr = \int_0^{\infty} \rho u 2\pi \, r dr, \qquad (8.9)$$

$$u_2 = (1 - 2a)u_0, \qquad (8.10)$$

where r_2 is the radial distance from the wake centerline at downstream distance x_1, ρ is the density of incompressible fluid (air), and u_2 is the wind speed just after the rotor. According to actuator disk theory, a is the induction factor. Substituting (8.10) and (8.7) into (8.9), we get

$$A(x) = \frac{a}{(\sigma/r_2^2)}, \qquad (8.11)$$

where σ is a linear function of x, and its estimation [29] is

$$\sigma = \frac{r}{\sqrt{2}} = \frac{r_0 + \alpha x}{\sqrt{2}}. \qquad (8.12)$$

Thus Jensen's Gaussian distribution wake model is given as

$$u = u_0 \left(1 - \frac{a}{(\sigma/r_2^2)} e^{\frac{-r^2}{2\sigma^2}} \right). \qquad (8.13)$$

Further, Frandsen's Gaussian version of wake model assumes that the mass flow rate through the control tube $Y_1 Y_3$ is not constant. Based on the principle of conservation of momentum, we get

$$T = \int_0^{\infty} \rho(u_0 - u)2\pi \, r dr = \frac{1}{2}\rho A_0(u_0^2 - u^2), \qquad (8.14)$$

where T is the net thrust in the presence of wake flow speed u. Further, substituting (8.11) and (8.7) into (8.14), we get

$$A(x) = 1 - \sqrt{\left(1 - \frac{C_t}{2(\sigma/r_0)^2}\right)}, \quad \sigma = \frac{r}{2} = \frac{r_0 + \alpha x}{2}, \qquad (8.15)$$

where $A(x)$ is the maximum normalized velocity deficit caused in wakes, and σ is the wake width for a given downstream distance x. We now propose a bilateral Gaussian wake model based on Jensen's and Frandsen's Gaussian wake approaches, which can be mathematically expressed as

$$u = u_0\left(1 - Q(x, r)\right), \quad Q(x, r) = A_j(x, r)A_f(x, r), \qquad (8.16)$$

$$\sigma_j = \frac{r_0 + \alpha x}{\sqrt{2}}, \qquad \sigma_f = \frac{r_0 + \alpha x}{2}, \qquad (8.17)$$

$$A_j(x, r) = \frac{a}{(\sigma_j/r_2^2)}e^{\frac{-r^2}{2\sigma_j^2}}, \quad A_f(x, r) = \left(1 - \sqrt{\left(1 - \frac{C_t}{2(\sigma_f/r_0)^2}\right)}\right)e^{\frac{-r^2}{2\sigma_f^2}}, \quad (8.18)$$

where $A_j(x, r)$ and $A_f(x, r)$ are the respective Jensen and Frandsen Gaussian components for a given downstream distance x and radial distance r, which is a function of the wake expansion factor α and downstream distance x. Further, σ_j and σ_f are the wake widths as function of x for Jensen's and Frandsen's models, respectively.

The motivation behind multiplying Jensen's ($A_j(x, r)$) and Frandsen's ($A_f(x, r)$) components comes from the individual advantages of the two models. Jensen's model is time-saving, whereas Frandsen's model holds good approximation for far-wake region $5D_0$–$7D_0$. The proposed bilateral Gaussian wake model was tested against benchmark models like Jensen's and Frandsen's wake models for single-wake and multiple-wake scenarios, and its results are presented next. In the case of our proposed model the wake velocity u is calculated for a fixed radial distance for a given wake expansion factor α as listed in Table 8.1.

8.2.3 Case study for single-wake model

We discuss two analytical models, Jensen's and Frandsen's single wake models. To analyze the performance of these individual models, the wind speed data for a wind site WBZ (earlier known as Westinghouse Broadcasting, the radio station in Boston), Tower Hull in Boston Harbor, is collected throughout September 2006. The wind speed is measured every 10 minutes at a hub height of 61 meters by a cup anemometer with an accuracy of

Table 8.1 Turbine specifications for wake model calculation.

Parameter	Value
Rotor radius (r_0)	38.5 m
Thrust coefficient (C_t)	0.88
Wake constant (α)	0.05
Hub height (h)	61 m

$\pm 2\%$. The wind speed series chosen for 250 data points has maximum and minimum wind speed of 13.05 m/s and 1.5 m/s, respectively, and a mean wind speed of 7.58 m/s.

The wake growth constant α is calculated using empirical relationship (8.2) for given hub height h and surface roughness length z_0. Since the wake effect is analyzed in neutral atmospheric boundary layer, the atmospheric stability correction factor $\psi_m(h/L)$ is considered zero. For C_t, the value 0.88 is chosen based on the assumption that every turbine in a wind farm is operating at Betz limit, where the coefficient of power is 0.58, which occurs at the induction factor $a = 0.33$. For a neutral atmospheric boundary layer (ABL) operation, the value of C_t is found to be higher for increased power output [30]. The wake flow behind the upwind turbine depends on the wake growth constant α, which in turn depends on topographical features of the wind farm land.

Consider a single wake wind turbine layout in Fig. 8.2. The wind turbine WT_1 is the upwind turbine, and WT_2 is the downwind turbine. The wind speed for WBZ tower Hull varies in the range from 1.5 m/s to 13.0 m/s, given that the ABL is not stratified (the density variation of air in vertical direction is zero), the pressure drop at hub of each wind turbine in the farm is constant, and the thrust coefficient (C_t) remains constant for wake stream flow. The effective wind speed observed is simulated using (8.3) and (8.4), and RMSE and the coefficient of determination R^2 were calculated to assess the model performance for a single wake model. Table 8.2 depicts the performance metrics, and Frandsen's single wake model outperforms Jensen's model in terms of RMSE and R^2.

The proposed and benchmark models already described earlier are tested for different downwind distances ($x = 2.5D_0$, $3D_0$, and $5D_0$). Fig. 8.4 shows the effective wind speed observed at wind turbine WT_2 due to the wake effect of WT_1 for Jensen', Frandsen's, and the proposed bilateral Gaussian wake models for a downwind distance of $2.5D_0$, and u_0 refers to the freestream wind speed.

Table 8.2 Performance indices for single-wake scenario.

Model	Metric	$2.5D_0$	$3D_0$	$5D_0$
Jensen	RMSE	3.7865	3.5461	2.7769
	R^2	0.8628	0.8683	0.8605
Frandsen	RMSE	1.0668	0.9375	2.7769
	R^2	0.8885	0.8780	0.8964
Proposed	RMSE	0.7148	0.7146	0.7169
	R^2	0.8976	0.8823	0.8971

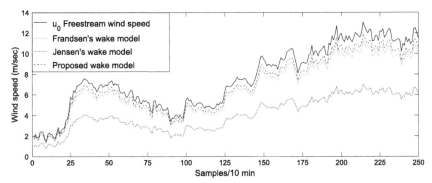

Figure 8.4 Wake effect on WT_2 based on the proposed and benchmark models.

Based on the RMSE, the proposed bilateral Gaussian wake model outperforms Jensen's model by 81.11% and Frandsen's model by 32.99% for a downstream distance of $2.5D_0$.

8.2.4 Multiple-wake model

This study can be further extended for wind wakes due to multiple upwind turbines. Due to shadowing effect created by wind turbines placed at different locations, the effective wind speed in the presence of wakes using Jensen's model [31] is

$$u = u_0\left(1 - \sum_{i=1}^{N}\left(1 - \sqrt{1 - C_t}\left(\frac{r_0}{r_{ij}}\right)^2 \frac{A_{sh,i}}{A_0}\right)\right), \qquad (8.19)$$

where $A_{sh,i}$ is the overlap area experienced by wind turbine under shadow from upwind turbine, and

$$A_{sh,i} = r_0^2\cos^{-1}\left(\frac{d_{ij}^2 + r_0^2 - r_{ij}^2}{2d_{ij}r_0}\right) + r_{ij}^2\cos^{-1}\left(\frac{d_{ij}^2 + r_{ij}^2 - r_0^2}{2d_{ij}r_{ij}}\right)$$

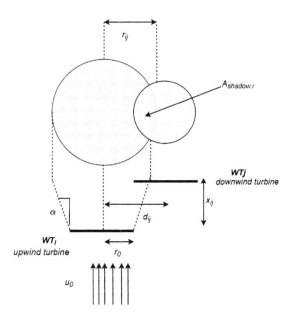

Figure 8.5 Shadowing effect in case of multiple wake scenario.

$$-\frac{1}{2}\sqrt{\left((r_0 + d_{ij})^2 - r_{ij}^2\right)\left(r_{ij}^2 - (r_0 - d_{ij})^2\right)}, \tag{8.20}$$

and based on Frandsen's wake model, the effective wind speed during multiple-wake scenario is given as

$$u = u_0\left(\frac{1}{2} \pm \frac{1}{2}\sqrt{1 - 2C_t\left(\frac{r_0}{r_{ij}}\right)^2\left(\frac{A_{sh,i}}{A_0}\right)}\right), \tag{8.21}$$

where d_{ij}, r_0, r_{ij} are the horizontal distances between upwind turbine WT_i and downwind turbine WT_j, the rotor radius for all turbines in wind farm, and the radius of the downwind turbine WT_j due to the wake effect of WT_i, and N is the total number of wind turbines causing shadowing effect.

Fig. 8.5 shows the overlap area experienced by a downwind turbine. Based on the overlap area, the degree of shadowing is calculated for multiple-wake scenario, where a wind turbine receives wind wakes from more than one upwind turbines. For our proposed bilateral Gaussian wake model, the effective wind speed at WT_j due to N upwind turbines WT_i

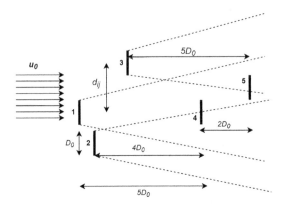

Figure 8.6 Wind farm layout consisting five wind turbines.

$(i = 1, 2, \ldots, N)$ is given as

$$u = u_0 \left(1 - \sum_{i=1}^{N} Q_{ij}(x, r) \right),$$ (8.22)

$$Q_{ij}(x, r) = A_{ij}(j) A_{ij}(f) D_{o,i}, \quad D_{o,i} = \left(\frac{A_{sh,i}}{A_0} \right),$$ (8.23)

where N is the total number of upwind turbines causing shadowing on the downwind turbine WT_j, $D_{o,i}$ is the degree of overlap due to shadowing, and $A_{ij}(j)$ and $A_{ij}(f)$ are the Jensen and Frandsen components calculated using (8.18). To test Jensen's and Frandsen's models for multiple-wake condition, a five-turbine wind farm layout (WFL) is chosen in Fig. 8.6. To study the multiple wake effect, the following assumptions are made:

A1 The wind turbines in the wind farm have same rotor diameter D_0.

A2 After a downwind distance of $10D_0$–$12D_0$, the wake effect due to wind turbine WT_1 disappears.

The effective wind speeds observed at WT_4 and WT_5 due to multiple upwind turbines are calculated using (8.19) and (8.21). The effective wind speeds at WT_4 and WT_5 due to the multiple wake effect are shown in Fig. 8.7.

We observe that Frandsen's model for multiple wake outperforms Jensen's model. Further, as the distance x_{ij} between upwind and downwind turbines increases, the wake effect diminishes. The performance of Jensen's and Frandsen's models multiple wake conditions is tested for wind turbines WT_4 and WT_5 for a five-turbine wind farm layout. See Fig. 8.8.

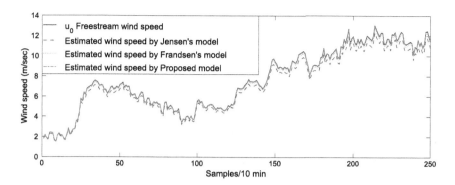

Figure 8.7 Effective wind speed at WT_4 due to wake effect of WT_1 and WT_2.

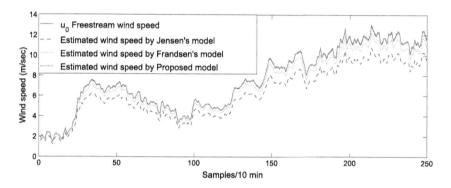

Figure 8.8 Effective wind speed at WT_5 due to wake effect of WT_1, WT_3 and WT_4.

The performance metric RMSE for the benchmark models and proposed model is evaluated. For WT_4, the RMSE is found to be 46.70% using Jensen's model and 15.88% using Frandsen's model. Similarly, for WT_5, the RMSE based on Jensen's model is 142.91%, and based on Frandsen's model, it is 50.81%. Compared with benchmark models, the RMSE values based on the proposed model for WT_4 and WT_5 are found to be 3.85% and 7.56%, respectively. So the velocity distribution at the far-wake end well complies with experimental results [32].

Further, the proposed wake model is validated by testing for multiple-wake scenarios for a dataset D2. Wind speed data are hourly measured for a wind farm Sotavento, Spain, in March 1–30, 2018, by a cup anemometer placed at a height of 61 meters. The first 250 data points are chosen for testing multiple wake scenario. The descriptive statistics for dataset D2 are $u_{max} = 17.3$ m/s, $u_{min} = 0.35$ m/s, $u_{mean} = 7.625$ m/s, and the standard

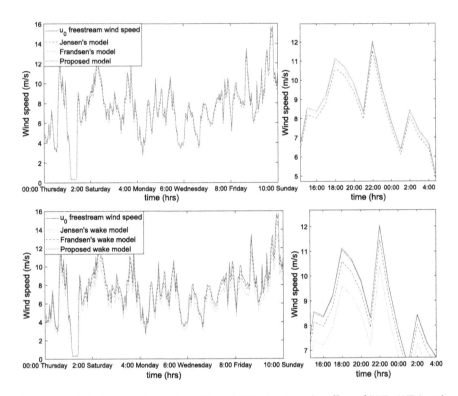

Figure 8.9 Effective wind speed at WT_4 and WT_5 due to wake effect of (WT_1, WT_2) and $(WT_1, WT_3,$ and $WT_4)$ for dataset D2.

deviation of 2.677. For wind turbine WT_4, RMSE is found to be 33.87% using Jensen's model, 11.72% using Frandsen's model, and 2.344% using the proposed model. For wind turbine WT_5, the RMSE for Jensen's model is 111.44% and 39.2% using Frandsen's model. Compared to the benchmark model, the proposed model recorded RMSE of 4.2%. Fig. 8.9 shows the wake velocities for WT_4 and WT_5.

8.3 Wake effect in wind forecasting

Wind forecasting is an essential step in terms of planning and operating a wind farm that is tied to a utility grid. Various methods based on time-series method like ARIMA and intelligent learning algorithms like ANNs and SVR are preferred for wind forecasting [33]. We study the forecasting under the presence of wind wakes. In a particular wind farm, not all wind turbines receive the wind speed at same magnitude and direction. Also, for a multiple wake condition, a wind turbine experiences wake due to multiple upwind turbines, but not all wind turbines cause significant wake effect and turbulence on the downwind turbine. Since not all upwind turbines cause velocity deficit, to identify the significant upwind turbines on a particular downwind turbine, grey correlation analysis (GRA) is performed. Next, we discuss various steps involved in GRA.

Grey correlation analysis

Grey correlation analysis or GRA is an important financial tool often used in system analysis technique [34]. The central idea of GRA is to establish the degree of closeness among various decision making sequences and a reference sequence. Other criteria affecting decision making apart from wake affect have been considered elsewhere [35]. Forecasting accuracy considering wake effects from multiple upwind turbines is greatly affected by the results of GRA.

Further, once the GRA is done, only the variables that have higher grey correlation degrees are chosen as inputs for the wind speed forecasting model. Grey correlation analysis is done as follows:

1. A reference sequence with the same length as that of decision making sequences is chosen as $X_0 = (x_{01}, x_{02}, \ldots, x_{0n})$, and a decision sequence is expressed as a matrix L representing all the decision sequences along with the reference sequence X_0, $X_i = (x_{i1}, x_{i2}, \ldots, x_{in})$, where $i = 1, 2, \ldots, m$:

$$L = \begin{bmatrix} x_{01} & x_{11} & \cdots & x_{m1} \\ x_{02} & x_{12} & \cdots & x_{m2} \\ \vdots & \vdots & \ddots & \vdots \\ x_{0n} & x_{1n} & \cdots & x_{mn} \end{bmatrix}. \tag{8.24}$$

2. Next, all the sequences are standardized either through transforms such as initial value transform, average value transform, or polar difference transform. The matrix L is then transformed as

$$L' = \begin{bmatrix} x'_{01} & x'_{11} & \cdots & x'_{m1} \\ x'_{02} & x'_{12} & \cdots' & x'_{m2} \\ \vdots & \vdots & \ddots & \vdots \\ x'_{0n} & x'_{1n} & \cdots & x'_{mn} \end{bmatrix}. \tag{8.25}$$

3. After standardization, the absolute differences of the corresponding elements of reference sequence and decision sequence are calculated, that is, $\Delta_{ik} = |x'_{0t} - x'_{ik}|$, $i = 1, 2, \ldots, n$, $k = 1, 2, \ldots, m$.
4. $gg = \min(\Delta_{ik})$ and $hh = \max(\Delta_{ik})$ are calculated.
5. The relational coefficient between the reference sequence x'_{0t} and decision sequence x'_{it} are calculated using

$$r(x'_{0t}, x'_{it}) = \frac{gg + \rho \times hh}{\Delta_{it} + \rho \times hh}, \tag{8.26}$$

where $\rho \in (0, 1)$ is called the distinguish coefficient. Usually, $\rho = 0.5$ is taken for all the calculations.
6. The grey relational degrees

$$r(X_0, X_i) = \frac{1}{n} \sum_{i=1}^{n} r(x'_{0t}, x'_{it}), \quad i = 1, 2, \ldots, m, \tag{8.27}$$

are calculated and ordered in descending order. Then the first jth inputs are selected to forecast wind speed in the presence of wakes.

In this study, GRA is used to identify the upwind turbines that significantly affect the given downwind turbine. The wake wind speed due to an individual upwind turbine is treated as the decision sequence and freestream wind speed u_0 are treated as the reference sequence X_0. The short-term wind speed forecasting is carried without and without GRA, and the forecasting results in terms of RMSE are compared for the proposed bilateral Gaussian wake model and benchmark models.

8.4 Results

This section throws light on the framework for short-term forecasting in the presence of wind wakes for two different wind farm layouts, that is, five-turbine wind farm layout (Fig. 8.4) and 15-turbine wind farm layout (Fig. 8.13). Wind speed data are collected for two data sets, a wind farm WBZ tower Hull, Boston Harbor, Massachusetts, and another one located in Sotavento, Galicia, Spain. Table 8.3 shows the descriptive statistics for the two wind farms with their sampling time.

Fig. 8.10 shows the wind speed time series for the two wind farms WBZ tower Hull and Sotavento, Galicia, Spain. These two datasets D1 (500 data points) and D2 (720 data points) are obtained at every 10 minutes and every hour, respectively, in the period of March 1–30, 2018.

Table 8.3 Descriptive statistics for wind speed for WBZ tower Hull, MA, and Sotavento, Spain.

Wind Farm	Dataset	Max (m/s)	Min (m/s)	Mean (m/s)	Std Dev
WBZ tower, MA	D1	10.14	0.36	4.4966	2.3143
Sotavento, Spain	D2	17.53	0.35	7.1768	2.8009

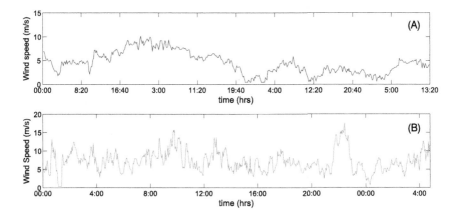

Figure 8.10 Wind speed time-series: (A) WBZ tower Hull, MA; (B) Sotavento, Spain.

The short-term wind speed forecasting, that is, 10 minutes for D1 and 1 hour for D2, is analyzed on the basis of standard metrics like RMSE, the sum of squared residuals, and the sum of squared deviations of testing samples (SST). Mathematical expressions for these metrics are

$$\text{RMSE} = \sqrt{\frac{1}{N}\sum_{i=1}^{N}(\hat{X}_i - X_i)^2} \times 100\%, \quad \text{SSR/SST} = \frac{\sum_{i=1}^{N}(\hat{X}_i - \bar{X})^2}{\sum_{i=1}^{N}(X_i - \bar{X}_i)^2},$$

$$(8.28)$$

where $\hat{X}_i$, X_i, and $\bar{X}$ are the predicted, actual, and mean values of the N testing samples. Lower the RMSE and value better the forecasting model. A high SSR/SST ratio indicates a good agreement between actual and estimated values of testing samples.

8.4.1 Forecasting results for five-turbine wind farm layout

The forecasting process involves first calculating the wind speed received by the downwind turbine under wake effect from individual upwind tur-

Table 8.4 Grey correlation degree and its ranking for five-turbine layout for datasets D1 and D2.

Grey correlation degree	D1		D2	
	Value	Rank	Value	Rank
$r(u_0, u_{15})$	0.7074	3	0.5867	3
$r(u_0, u_{35})$	0.8529	1	0.6748	1
$r(u_0, u_{45})$	0.8234	2	0.5940	2

bines. The wind farm layouts for five-turbine wind farm and 15-turbine farm are selected. In the case of wind farm layout with five turbines, wind farms WT_4 and WT_5 are the downwind turbines. The wake flow speeds for both wind turbines are calculated using Jensen's and Frandsen's models. The turbine parameters for both wind farm layouts are chosen the same for simplicity and are listed in Table 8.1.

Wind turbine WT_4 experiences wake from WT_1 and WT_2, whereas wind WT_5 experiences wake from WT_1, WT_3, and WT_4. The distance x_{ij} between upwind WT_i and downwind turbines WT_j has already been listed in Section 8.2.4. After calculation of wake flow speed using (8.3) and (8.4), the wake flow speed time series is analyzed for grey correlation analysis by comparing it with a reference sequence X_0 (here the freestream wind speed u_0). The grey correlation degree (GCD) $r(u_0, u_{ij})$ obtained thereafter is arranged in descending order of its magnitude, and the first jth inputs are selected. In the case of five-turbine layout the GCD was found for wind speed series u_{15}, u_{35}, and u_{45} to select inputs for forecasting wind speed for turbine WT_5. The GRA results for u_{15}, u_{35}, and u_{45} are highlighted in Table 8.4.

The GCDs in descending order are $r(u_0, u_{35}) > r(u_0, u_{45}) > r(u_0, u_{15})$, where u_{35} and u_{45} are the inputs to the SVR model to forecast the wind speed for WT_5. The wind speed for WT_5 is also forecasted without considering GRA by selecting all the wake causing wind turbines as input to the SVR model. The effective wind speed is decomposed into approximate signal (a5) and detail signals (d1, d2, d3, d4, d5) using five-level Daubechies 4 (db4) wavelet transform.

The forecasting is done using SVR model where the data are divided as the training set (first 80%) and testing set (remaining 20%). The SVR model uses the radial basis function (RBF) as a kernel function. The SVR hyperparameters are chosen from the set $\{2^i, i = -9, -8, \ldots, 10\}$. This choice of the search space enables us to determine optimal parameters from a finite set of real numbers and leads to fast computation. We selected the RBF

Table 8.5 Forecasting results for wind turbine WT_5 for datasets D1 and D2.

Dataset	Metric	Jensen's model	
		without GRA	with GRA
D1	RMSE (%)	6.53	5.84
	SSR/SST	1.0897	1.0124
D2	RMSE (%)	12.93	11.86
	SSR/SST	1.0483	1.0488
		Frandsen's model	
		without GRA	with GRA
D1	RMSE (%)	3.68	3.31
	SSR/SST	1.0986	1.0928
D2	RMSE (%)	9.02	6.43
	SSR/SST	1.0552	1.0578
		Proposed model	
		without GRA	with GRA
D1	RMSE (%)	2.86	2.56
	SSR/SST	1.1087	1.1902
D2	RMSE (%)	7.33	4.91
	SSR/SST	1.0270	1.0020

kernel function for SVR forecasting, the bandwidth was chosen as $\sigma = 2^5$, and the regularization constant γ was taken as 2^2 [36]. Table 8.5 shows the forecasting results for wind turbine WT_5 based on Jensen's and Frandsen's wake models for the two datasets D1 and D2.

The wind forecast accuracy for Frandsen's model outperforms Jensen's model. Using our proposed model, short-term wind forecasting is carried out for datasets D1 and D2, and the results are studied with respect to benchmark models. For dataset D1, the RMSE is found to be less than in Jensen's and Frandsen's models. For dataset D1, without GRA, the RMSE is 2.86%, and for dataset D2, the RMSE is 7.33%. For dataset D1, the RMSE value without GRA-based Jensen's model is 6.53%, and based on Frandsen's model, it is 3.68%. The SSR/SST ratio for Jensen's model is 1.0897, and using Frandsen's model, it is 1.0986. By incorporating GRA in our forecasting model, the RMSE for Frandsen's model is 3.31%, and for Jensen's model, it is 5.84%. The SSR/SST ratio for Frandsen's model is found to be better than for Jensen's model, thus suggesting better wind speed forecasting based on this model. A high SSR/SST ratio implies a good agreement between actual and estimated values.

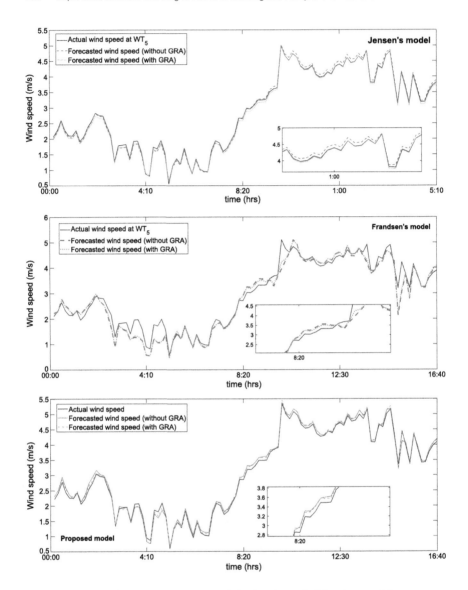

Figure 8.11 Wind forecasting for WT_5 with and without GRA for dataset D1.

Figs. 8.11 and 8.12 show the short-term wind speed forecasting results for WT_5 in the presence of wakes based on Jensen's, Frandsen's, and the proposed model for datasets D1 and D2, respectively.

The short-term forecast was done using the hybrid method wavelet-SVR for which 80% (400 data points) of data were used for training, and

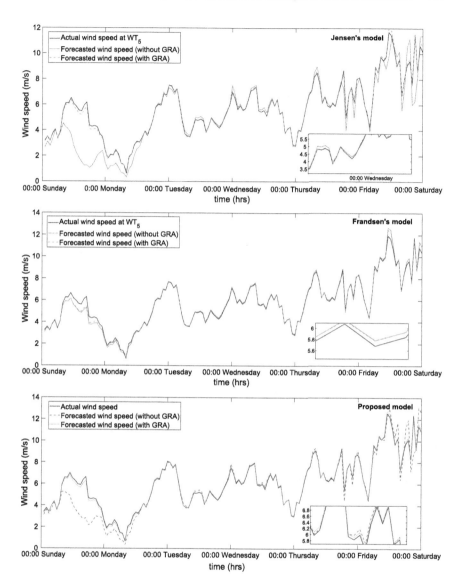

Figure 8.12 Wind forecasting for WT_5 with and without GRA for dataset D2.

20% (100 data points) were used for testing. For dataset D2, the effective wind speed for wind turbine WT_5 is calculated based on Jensen's and Frandsen's models, and GRA is done for inputs u_{15}, u_{35}, and u_{45}. Based on the GCD rankings, we select u_{35} as the input to the SVR model along with decomposition signals using the wavelet transform. The RMSE values for

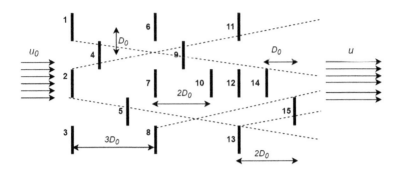

Figure 8.13 Schematic for a 15-turbine wind farm layout.

dataset D2 without GRA based on Jensen's model is 12.93%, and based on Frandsen's model, it is 9.02%. The SSR/SST ratio for Jensen's model is 1.0483, and using Frandsen's model, it is 1.0552. A high SSR/SST ratio implies that the regressor extracts maximum statistical information from the data. A wake model estimating lower velocity deficit yields higher annual electricity production (AEP).

8.4.2 Forecasting results for 15-turbine wind farm

Next, we discuss short-term wind speed forecasting in the presence of wind wakes for a 15-turbine wind farm layout with turbines arranged asymmetrically as in Fig. 8.13.

The wind turbine WT_{12} is affected from wakes by upwind turbines WT_3, WT_4, WT_5, WT_7, WT_9, and WT_{10}. The effective wind speed due to upwind turbines is calculated using (8.19) and (8.21) for Jensen's and Frandsen's models, respectively:

$$u_{12eff} = u_0\left(1 - \sum_{i=1}^{N}\left(1 - \sqrt{1 - C_t}\left(\frac{r_0}{r_{ij}}\right)^2\frac{A_{sh,i}}{A_0}\right)\right), \qquad (8.29)$$

where u_{12eff} is the effective wind speed at WT_{12} due to wake effects of upwind turbines. The overlap area $A_{sh,i}$ due to upwind turbine WT_i is found using (8.20). The wind speeds $u_{i,j}$ due to WT_i, $i = 3, 4, 5, 7, 9, 10$, based on Jensen's, Frandsen's, and the proposed model, are calculated. In Table 8.6, we calculate the grey correlation degree $r(u_0, u_{i,j})$, with u_0 (freestream velocity) as reference, using GRA for datasets D1 and D2.

The SVR model is trained for 80% of the data and tested for the remaining 20% of data. The RBF kernel function with bandwidth $\sigma = 2^5$ is

Table 8.6 Grey correlation degree and ranking for 12-turbine layout for D1 and D2.

Grey correlation degree	D1		D2	
	Value	Rank	Value	Rank
$r(u_0, u_{3,12})$	0.8218	3	0.6644	5
$r(u_0, u_{4,12})$	0.8529	1	0.6748	4
$r(u_0, u_{5,12})$	0.7809	5	0.7065	3
$r(u_0, u_{7,12})$	0.7228	6	0.8246	1
$r(u_0, u_{9,12})$	0.8234	2	0.5940	6
$r(u_0, u_{10,12})$	0.7917	4	0.7260	2

Table 8.7 Forecasting results for wind turbine WT_{12} for datasets D1 and D2.

Dataset	Metric	Jensen's model	
		without GRA	with GRA
D1	RMSE (%)	5.86	4.66
	SSR/SST	0.9866	1.0224
D2	RMSE (%)	8.06	6.86
	SSR/SST	1.0519	1.2934

		Frandsen's model	
		without GRA	with GRA
D1	RMSE (%)	4.78	4.38
	SSR/SST	1.0933	1.0818
D2	RMSE (%)	6.38	6.21
	SSR/SST	1.0297	1.3603

		Proposed model	
		without GRA	with GRA
D1	RMSE (%)	4.82	2.56
	SSR/SST	1.0599	1.0190
D2	RMSE (%)	5.52	4.72
	SSR/SST	1.0363	1.0150

used. For dataset D1, the inputs to the SVR model are wavelet decomposition signals of effective wind speed u_{12eff} (a5; d1, d2, d3, d4, and d5) and wind speed series u_{ij} due to upwind turbines WT_i, $i = 3, 4, 5, 7, 9, 10$. The short-term wind speed is forecasted for WT_{12} once without GRA and once with GRA, that is, only selected inputs ($u_{5,12}$; $u_{7,12}$; $u_{10,12}$) based on their rankings are used to forecast the effective input at WT_{12}. Table 8.7 shows the performance metrics based on Jensen's, Frandsen's, and the proposed model.

For dataset D1, the RMSE for short-term wind forecasting without GRA is 5.86% with Jensen's model and 4.78% with Frandsen's model. With GRA, the RMSE for Jensen's model is 4.66%, and for Frandsen's model, it is 4.38%. The SSR/SST ratio was found better when wind forecasting in the presence of wakes is done with GRA than without it. The performance metrics indicate that wake modeling and wind forecasting based on Frandsen's model outperformed Jensen's model. Figs. 8.14 and 8.15 show the forecasting results for wind turbine WT_{12} based on Jensen's, Frandsen's, and the proposed models for datasets D1 and D2.

Similarly, for dataset D2, the RMSE without GRA for Jensen's model is 8.06%, whereas for Frandsen's model, it is 6.38%. With GRA, the RMSE for Jensen's model is 6.86%, whereas for Frandsen's model, it is 6.21%. It can be seen that RMSE for Frandsen's model is smaller than for Jensen's model, thus implying a better forecast in the presence of wind wakes for turbine WT_{12}. The SSR/SST ratio is consistent for Jensen's and Frandsen's models and shows good agreement between actual and estimated values.

Based on our proposed model, we find that GRA provides significantly better forecasting performance in terms of RMSE. The RMSE for dataset D1 is 4.82% without GRA and 2.56% with GRA. Similarly, for dataset 2, the RMSE is 5.52% without GRA and 4.72% with GRA. For a 15-turbine wind farm layout, the wind forecasting was carried out for WT_{12}, and we found that our proposed model yields better forecasting performance in the presence of wind wakes.

This chapter discusses the short-term wind forecasting in the presence of wakes by considering two different wind farm layouts and numbers of wind turbines. The wake effect is studied based on the models proposed by Jensen and Frandsen. For single-wake scenario, the proposed bilateral Gaussian wake model is compared with Frandsen's and Jensen's models for different downwind distances. In a wind farm a downwind turbine experiences wakes due to multiple turbines, thus creating a shadowing effect. The multiple-wake scenario is tested for a five-turbine wind farm layout, and our proposed model outperformed Frandsen's model by 75.75% and Jensen's model by 91.75% in terms of RMSE. The short-term wind forecasting is carried out using a hybrid method based on wavelet decomposition and SVR. The inputs for the forecasting model are selected based on grey correlation degree and only significant inputs, that is, upwind turbines are selected as inputs to forecast wind speed for the downwind turbine.

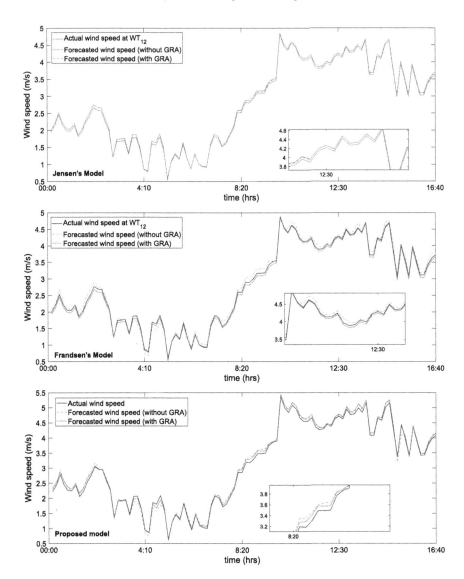

Figure 8.14 Wind forecasting for WT_{12} with and without GRA for dataset D1.

Based on our study, we find that for both wind farm layouts, with GRA, our proposed model outperforms Frandsen's and Jensen's models in terms of RMSE, and choosing only significant upwind turbines as input to the SVR forecasting model, we get much better forecast accuracy. The significance of GRA in identifying wind turbines that actually cause velocity deficit leads to an efficient micrositing of wind farms. GRA is

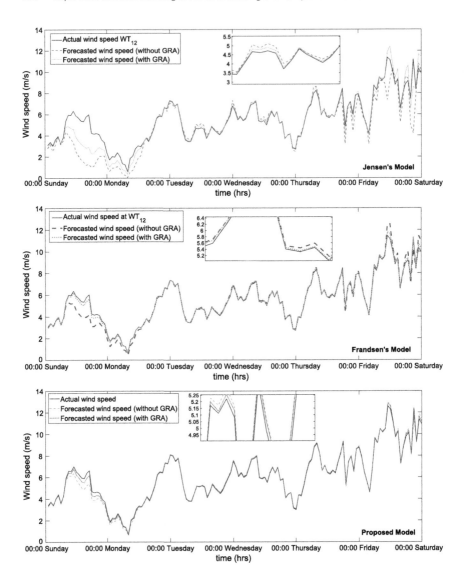

Figure 8.15 Wind forecasting for WT_{12} with and without GRA for dataset D2.

used to select input features for an SVR-based prediction model. SVR-based forecasting results in overestimating the wind speed predictions as indicated by SSR/SST index greater than 1. The overfitting scenario of SVR can be possibly overcome by using improved variants of SVR like least square support vector regression (LSSVR) and twin support vector regression (TSVR).

A GRA-based framework is used to study wind forecasting in the presence of wind wakes. Benchmark wake models like Jensen's and Frandsen's models are tested against a novel bilateral Gaussian wake model. The proposed wake model is based on Gaussian variation of Jensen's and Frandsen's models. The model is tested for two datasets D1 and D2 considering single-wake and multiple-wake scenarios. The proposed model outperforms Jensen's and Frandsen's models. Further, forecasting is carried out where the wind speed time-series is decomposed into approximate and detail signals using the Daubechies wavelet with a five-level decomposition. The wavelet transform removes the noise components present in the series, and SVR is used to forecast wind speed. GRA is used as an important tool to identify the upstream turbines significantly contributing to wake effect. Results reveal that for both wind farm layouts, GRA improves the forecast accuracy.

References

[1] S. Chowdhury, J. Zhang, A. Messac, L. Castillo, Unrestricted wind farm layout optimization (UWFLO): investigating key factors influencing the maximum power generation, Renewable Energy 38 (1) (2012) 16–30.

[2] P. Patel, A. Shandilya, D. Deb, Optimized hybrid wind power generation with forecasting algorithms and battery life considerations, in: 2017 IEEE Power and Energy Conference at Illinois (PECI), IEEE, 2017.

[3] J.F. Manwell, Wind Energy Explained: Theory, Design and Application, Wiley, Chichester, U.K., 2009.

[4] C.L. Archer, A. Vasel-Be-Hagh, C. Yan, S. Wu, Y. Pan, J.F. Brodie, A.E. Maguire, Review and evaluation of wake loss models for wind energy applications, Applied Energy 226 (2018) 1187–1207.

[5] P. Hashemi-Tari, K. Siddiqui, M. Refan, H. Hangan, Wind tunnel investigation of the near-wake flow dynamics of a horizontal axis wind turbine, Journal of Physics. Conference Series 524 (2014) 012176.

[6] J. Ainslie, Calculating the flowfield in the wake of wind turbines, Journal of Wind Engineering and Industrial Aerodynamics 27 (1–3) (1988) 213–224.

[7] G.C. Larsen, H.A. Madsen, K. Thomsen, T.J. Larsen, Wake meandering: a pragmatic approach, Wind Energy 11 (4) (2008) 377–395.

[8] S. Frandsen, R. Barthelmie, S. Pryor, O. Rathmann, S. Larsen, J. Højstrup, M. Thøgersen, Analytical modelling of wind speed deficit in large offshore wind farms, Wind Energy 9 (1–2) (2006) 39–53.

[9] H. Sun, H. Yang, Study on an innovative three-dimensional wind turbine wake model, Applied Energy 226 (2018) 483–493.

[10] R.J. Barthelmie, G.C. Larsen, S.T. Frandsen, L. Folkerts, K. Rados, S.C. Pryor, B. Lange, G. Schepers, Comparison of wake model simulations with offshore wind turbine wake profiles measured by sodar, Journal of Atmospheric and Oceanic Technology 23 (7) (2006) 888–901.

[11] L. Tian, W. Zhu, W. Shen, N. Zhao, Z. Shen, Development and validation of a new two-dimensional wake model for wind turbine wakes, Journal of Wind Engineering and Industrial Aerodynamics 137 (2015) 90–99.

[12] T. Ishihara, G.-W. Qian, A new Gaussian-based analytical wake model for wind turbines considering ambient turbulence intensities and thrust coefficient effects, Journal of Wind Engineering and Industrial Aerodynamics 177 (2018) 275–292.

[13] T. Göçmen, P. van der Laan, P.-E. Réthoré, A.P. Diaz, G.C. Larsen, S. Ott, Wind turbine wake models developed at the Technical University of Denmark: a review, Renewable & Sustainable Energy Reviews 60 (2016) 752–769.

[14] J.K. Sethi, D. Deb, M. Malakar, Modeling of a wind turbine farm in presence of wake interactions, in: 2011 International Conference on Energy, Automation and Signal, IEEE, 2011.

[15] H. Schümann, F. Pierella, L. Sætran, Experimental investigation of wind turbine wakes in the wind tunnel, Energy Procedia 35 (2013) 285–296.

[16] A. Crespo, J. Hernández, S. Frandsen, Survey of modelling methods for wind turbine wakes and wind farms, Wind Energy 2 (1) (1999) 1–24.

[17] R.J. Barthelmie, K. Hansen, S.T. Frandsen, O. Rathmann, J.G. Schepers, W. Schlez, J. Phillips, K. Rados, A. Zervos, E.S. Politis, P.K. Chaviaropoulos, Modelling and measuring flow and wind turbine wakes in large wind farms offshore, Wind Energy 12 (5) (2009) 431–444.

[18] F. Porté-Agel, Y.-T. Wu, C.-H. Chen, A numerical study of the effects of wind direction on turbine wakes and power losses in a large wind farm, Energies 6 (10) (2013) 5297–5313.

[19] J. Park, K.H. Law, Layout optimization for maximizing wind farm power production using sequential convex programming, Applied Energy 151 (2015) 320–334.

[20] L. Guan-yang, W. Hongzhao, L. Guanglei, C. Yamei, L. Hong-zheng, S. Yi, Security and stability analysis of wind farms integration into distribution network, IOP Conference Series: Materials Science and Engineering 199 (2017) 012102.

[21] S.S. Soman, H. Zareipour, O. Malik, P. Mandal, A review of wind power and wind speed forecasting methods with different time horizons, in: North American Power Symposium 2010, IEEE, 2010.

[22] I. Okumus, A. Dinler, Current status of wind energy forecasting and a hybrid method for hourly predictions, Energy Conversion and Management 123 (2016) 362–371.

[23] J. Zhou, J. Shi, G. Li, Fine tuning support vector machines for short-term wind speed forecasting, Energy Conversion and Management 52 (4) (2011) 1990–1998.

[24] P. Patel, D. Deb, Battery state of charge based algorithm for optimal wind farm power management, in: 2017 6th International Conference on Computer Applications in Electrical Engineering-Recent Advances (CERA), IEEE, 2017.

[25] N. Jensen, A Note on Wind Generator Interaction, 1983.

[26] R.B. Stull (Ed.), An Introduction to Boundary Layer Meteorology, Springer Netherlands, 1988.

[27] Y.-T. Wu, F. Porté-Agel, Atmospheric turbulence effects on wind-turbine wakes: an LES study, Energies 5 (12) (2012) 5340–5362.

[28] M. Bastankhah, F. Porté-Agel, A new analytical model for wind-turbine wakes, Renewable Energy 70 (2014) 116–123.

[29] L. Sizhuang, F. Youtong, Analysis of the Jensen's model, the Frandsen's model and their Gaussian variations, in: 2014 17th International Conference on Electrical Machines and Systems (ICEMS), IEEE, 2014.

[30] A. Sathe, J. Mann, T. Barlas, W. Bierbooms, G. van Bussel, Influence of atmospheric stability on wind turbine loads, Wind Energy 16 (7) (2012) 1013–1032.

[31] F. González-Longatt, P. Wall, V. Terzija, Wake effect in wind farm performance: steady-state and dynamic behavior, Renewable Energy 39 (1) (2012) 329–338.

[32] C.D. Markfort, W. Zhang, F. Porté-Agel, Turbulent flow and scalar transport through and over aligned and staggered wind farms, Journal of Turbulence 13 (2012) N33.

[33] Q. He, J. Wang, H. Lu, A hybrid system for short-term wind speed forecasting, Applied Energy 226 (2018) 756–771.

[34] S.-F. Yin, X. jing Wang, J. hui Wu, G. li Wang, Grey correlation analysis on the influential factors the hospital medical expenditure, in: Information Computing and Applications, Springer Berlin Heidelberg, 2010, pp. 73–78.

[35] H.S. Dhiman, D. Deb, V. Muresan, M.-L. Unguresan, Multi-criteria decision making approach for hybrid operation of wind farms, Symmetry 11 (5) (2019) 675, https://doi.org/10.3390/sym11050675.

[36] H. Duan, R. Wang, X. Liu, H. Liu, A method to determine the hyper-parameter range for tuning RBF support vector machines, in: 2010 International Conference on E-Product E-Service and E-Entertainment, IEEE, 2010.

Epilogue

Wind speed prediction is an integral part of the wind energy sector. The power available in wind resource is enormous and hence requires its accurate assessment for a given regime. A modern day power grid relies on the availability of reserve power capacity, and this dependency can escalate the system cost. Wind forecasting based on the future time horizon plays an important role in the deregulated market structure. From an economic perspective, wind power schedules available ahead in time can lead to optimal power dispatch for a large interconnected power system. With the concept of hybrid wind farms getting popular among developing and developed countries, the research in the field of forecasting has increased tremendously.

Wind speed forecasting in the modern day markets is carried out with a sole purpose to plan the reserve power capacity. An erratic forecast can jeopardize the system equilibrium in terms of generation and demand. To solve this issue of accurate wind power forecast, machine intelligent models in tandem with statistical models are used to improve short-term and medium-term forecasts. The nonlinearity in wind speed time series causes statistical models to fail, thus paving the way for machine intelligent models. Commonly used signal decomposition techniques that aim to address the volatility in wind speed time series aid the accuracy of short-term wind speed prediction. The wavelet transform and empirical mode decomposition are among the popular techniques that are implemented for decomposing wind speed time series. Apart from these wavelet packet decomposition technique that decomposes both low-frequency and high-frequency signals at each stage is also used. However, due to its time consumption issue, it is not preferred. In this book, models based on classical support vector regression and its variants like LSSVR, TSVR, and ε-TSVR are discussed with their performance under different real-time wind speed datasets.

Whereas these models belong to the category of supervised machine learning, we also discuss the decision tree ensemble models like random forest and gradient boosted machines for wind speed and ramp event prediction. We also discuss the ramp event prediction where a sudden change in wind speed magnitude is likely to affect the wind farm operation. Different wind speed datasets across the globe are taken and assessed for ramp prediction. The potential superiority of decision tree ensemble models over SVR models and its variants is also discussed where the problem of over-fitting causes large errors for raw wind speed data. We also discuss a special application of machine learning-based forecasting models for wind farms in the presence of wake interactions. Two case studies with different wind farm layouts are considered, and relevant wind turbines affecting prediction of a given turbine are identified.

Overall, an integrated solution that aims to find a best possible machine learning model for wind speed time series can be found for optimal wind power dispatch considering the reserve power capacity. Wind farm operators also face an important challenge of decision making, which has its roots from the forecasting technique incorporated to plan their wind power dispatch [1–3]. For a hybrid wind farm, operational strategies that a wind farm operator decides to take are also dependent on the reserve power capacity available. For ensuring a good balance between forecasts and reserve capacity, an integrated approach considering the wake effect that alters the power production of a wind farm can be deduced. Wake effects can be controlled up to a certain amount by implementing redirection strategies. These strategies involve accurate measurement devices like LIDAR, which gives an estimate of wind speed just before it actually interacts with the wind turbine. Availability of such devices helps a wind farm operator to take necessary steps in conjugation with the operational strategy to be implemented. Since wind markets require clearances at a short interval of 30 minutes to 1 hour, more focus is laid on short-term wind forecasting, which also tries to encompass weather models. The integrated methodology will determine the best forecasting technique, which under the effect of wake will lead to savings in reserve power capacity in terms of BESS and increase its life time.

References

[1] H.S. Dhiman, D. Deb, Decision and Control in Hybrid Wind Farms, Springer, 2019.
[2] H. Dhiman, D. Deb, V. Muresan, V. Balas, Wake management in wind farms: an adaptive control approach, Energies 12 (7) (2019) 1247, https://doi.org/10.3390/en12071247.
[3] H.S. Dhiman, D. Deb, V. Muresan, M.-L. Unguresan, Multi-criteria decision making approach for hybrid operation of wind farms, Symmetry 11 (5) (2019) 675, https://doi.org/10.3390/sym11050675.

APPENDIX A

Introduction to R for machine learning regression

A.1 Data handling in R

Data in R can be handled in several forms like vectors, lists, matrices, and data frames. Commonly used forms are vectors and lists as they are easy to work with. Vectors and matrices represent homogeneous data, whereas lists and data frames deal with heterogeneous data.

A.1.1 Data handling via vectors, lists, and data frames

Example A.1. Creating vectors in R

```
Player <- c("Cristiano","Messi","Neymar")
Salary<-c(35000,50000,100000)
```

Example A.2. Creating lists in R

```
Profile <- list(name=c("Cristiano","Messi","Neymar"),
salary=c(75000,50000,100000))
```

Output

```
Profile
$name
[1] "Cristiano" "Messi" "Neymar"
$salary
[1] 75000 50000 100000
```

A.1.2 Importing data through .xlsx and .csv files

Most of the data used in machine learning applications are stored in forms of column vectors, which are saved in .xlsx or .csv files. Data handling, storing, manipulation, and computation offered by Microsoft Excel file format is undisputed in terms of popularity and security. Conveniently, in R the Excel files with large chunk of data can be directly imported into workspace by using simple one line command. The following code shows how to import wind data along with several other weather variables for a site in Brazil; see the output of the data imported and visualize the summary of the imported data.

Example A.3. Example to import data from .xlsx file

```
library(xlsx)
read.xlsx("wind_data_brazil.xlsx",sheetName="Sheet2")
View(wind_data_brazil)
```

Output

Sr no.	Ozone	Solar R	Wind	Temp	Month	Day
1	41	190	7.4	67	5	1
2	36	118	8.0	72	5	2
3	12	149	12.6	74	5	3
4	18	313	11.5	62	5	4
5	NA	NA	14.3	56	5	5

Summary

```
summary(wind_data_brazil$Wind)
```

Min.	1st Qu.	Median	Mean	3rd Qu.	Max.
1.700	7.400	9.700	9.958	11.500	20.700

Importing .csv files works in a similar way. We further give an example to import a file named "Anholt_Denmark_March2019.csv". To view the summary of the variables, the same command can be used as in case of .xlsx files.

Example A.4. Example to import data from .csv files

```
library(readr)
data <- read_csv("Anholt_Denmark_March2019.csv")
View(data)
```

A.1.3 Line plots, histograms, and autocorrelation function for wind speed in R

Consider a list of variables f1, consisting of wind speed, temperature, humidity, and air pressure for a site. To plot the distribution of wind speed for this site, R commands can be conveniently used as follows.

Example A.5. R code for Line plot & Histogram

```
plot(f1$windspeed,xlab = "time",ylab = "Wind speed")
```

See Fig. A.1.

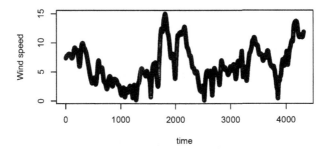

Figure A.1 Line plot for wind speed.

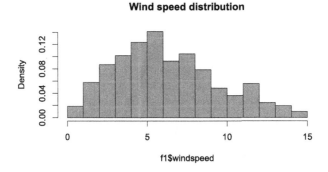

Figure A.2 Wind speed probability density distribution using R command.

```
hist(f1$windspeed,main="Wind speed distribution",col="red",
probability = TRUE)
```

See Fig. A.2.

We now discuss the R commands for autocorrelation functions for wind speed. An autocorrelation function depicts the dependence of an element with respect to elements of the series itself. For a wind speed time series, autocorrelation aids in selecting the lag orders for ARMA and ARIMA models. Dominant lag instants are helpful in identifying the behavioral pattern in wind speed. Mathematically, autocorrelation at lag instant k for a time series Y_t is expressed as

$$r_k = \frac{\sum_{t=1}^{N-k} \left(Y_t - \overline{Y}\right)\left(Y_{t+k} - \overline{Y}\right)}{\sum_{i=1}^{N} \left(Y_t - \overline{Y}\right)^2}. \tag{A.1}$$

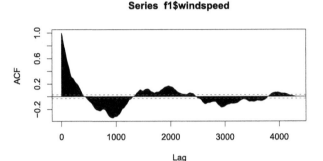

Figure A.3 ACF plot for time series f1$windspeed in R.

Example A.6. Autocorrelation of time-series

```
acf(f1$windspeed,lag.max = length(f1$windspeed),plot = TRUE)
```

See Fig. A.3.

A.1.4 Parameter estimation for wind speed probability density functions in R

Various PDFs like Weibull, Rayleigh, gamma, log-normal, and Lindley are based on parameters that decide the shape and scale of wind speed distributions. Estimation of such parameters is based on nonlinear minimization of loglikelihood function. In R the parameters can be estimated by applying defining functions for individual distributions. For example, the Stacy–Mihram model described in Chapter 2 has parameters α and β, which can be estimated by minimizing the negative of the loglikelihood function

$$L(\nu|\alpha, \beta) = n\log\beta - n\beta\log\alpha + (\beta - 1)\sum_{i=1}^{n}\log(\nu_i) - \sum_{i=1}^{n}\left(\frac{\nu_i}{\alpha}\right)^{\beta}. \quad (A.2)$$

Let us consider the parameter vector $\theta = (\alpha, \beta)$ such that $\theta[1] = \alpha$ and $\theta[2] = \beta$. The function `nlm` for nonlinear minimization can be used as follows.

Example A.7. Parameter estimation for distributions in R

```
n<-length(Y)
fn<-function(theta)
{
```

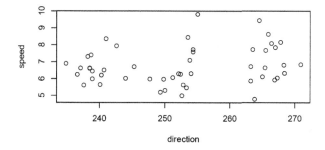

Figure A.4 Scatter plot between wind speed and direction.

```
-n*log(theta[2])+n*(theta[2])*log(theta[1])-(theta[2]-1)
*sum(log(Y))+sum((Y/theta[1])^theta[2])
}
theta_est<-nlm(fn, theta <- c(20,45), hessian=TRUE)
```

In these R commands, Y is a column vector depicting the wind speed time series, and n is the length of the time series. Note that the function fn minimized here is the negative of the loglikelihood function described by (A.2). The initial guess of the parameters is passed through a vector theta<-c(α_0, β_0). The estimates ($\hat{\alpha}, \hat{\beta}$) can be accessed using $ operator as theta_est$estimate.

A.2 Linear regression analysis in R

Linear regression aims to model a dependent variable Y as a function of independent variable X that follows an expression given as

$$Y = \phi_1 + \phi_2 X + v, \qquad (A.3)$$

where ϕ_1, ϕ_2, and v are the intercept, slope, and error terms for the regression model. Let us study how to carry out a simple linear regression in R. We take a data set Y as wind speed for a particular region and X as wind direction to see dependence of wind speed on wind direction and analyze error metrics. The data for wind speed and wind direction are collected from a wind farm Bishop & Clerks in USA. The scatter plot for 50 data points is illustrated in Fig. A.4. Further, Table A.1 depicts the first 10 data-points for wind speed and wind direction. The data can be imported in R workspace via .xlsx file as described in an earlier section.

Table A.1 Wind speed and direction data for linear regression.

Wind speed (Y) (m/s)	Wind direction (X) (degree)
6.51	240.7
6.63	237.1
7.31	238.2
6.45	238.9
6.21	240.3
5.99	238.9
5.64	240.1
5.97	249.8
6.27	252.4
5.46	253.3

Example A.8. R codes for linear regression

```
lm.fit<-lm(speed ~ direction)
summary(lm.fit)
```

Output

```
Call:
lm(formula = speed ~ direction)
Residuals:
Min         1Q        Median    3Q       Max
-2.1752   -0.7153   -0.2277   0.8084   3.0242
Coefficients:
              Estimate   Std. Error   t value   Pr(>|t|)
(Intercept)   1.87824    3.50950      0.535     0.595
direction     0.01925    0.01388      1.387     0.172
Residual standard error: 1.097 on 48 degrees of freedom
Multiple R-squared: 0.03852, Adjusted R-squared: 0.01849
F-statistic: 1.923 on 1 and 48 DF, p-value: 0.1719
```

From this R code output we can see that the wind speed is related to wind direction with a mathematical model given as (see Fig. A.5)

$$\mathrm{speed} = 1.87824 + 0.01925 \cdot \mathrm{direction}. \tag{A.4}$$

It is worth noting that the R^2 value for the current regression problem is very small. The predictions can be improved further if we use more than one independent variable.

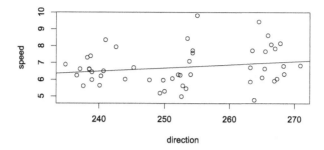

Figure A.5 Linear regression model for speed and distance in R.

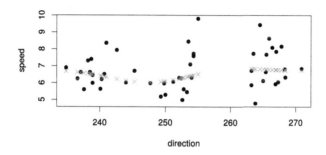

Figure A.6 Support vector regression model.

A.3 Support vector regression in R

Let us now see how to fit a support vector machine model for a regression model for the example studied in previous case. Here the data for wind speed and direction remain the same. We try to fit an svm model as described in Fig. A.6 and find out its error performance.

Example A.9. R codes for SVR

```
model <- svm(speed ~ direction)
predicted_speed <- predict(model, direction)
points(direction, predicted_speed, col = "red", pch=4)
rmse <- function(error)
{
sqrt(mean(error^2))
}
error <- model$residuals # same as speed-predicted_speed
predictionRMSE <- rmse(error)
```

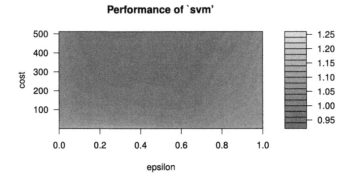

Figure A.7 Mean squared error after tuning SVM hyperparameters.

```
plot(direction,speed,pch=14)
points(direction,predicted_speed,col="red",pch=4)
```

SVM performance can be further improved by tuning the hyperparameters. A grid search can be performed in the range of 2^{-10}–2^{10} to determine optimal values of tolerance error ε and regularization parameter C. R commands for hyperparameters tuning are given next.

Example A.10. Hyperparameter tuning in R

```
tuneResult <- tune(svm, speed ~ direction, data = Data,
ranges = list(epsilon = seq(0,1,0.1), cost = 2^(2:9)) )
```

In this example, `Data` is a data frame consisting of wind speed and wind directions, which are column vectors. The results of hyperparameter tuning can be accessed using `print(tuneResult)`. See Fig. A.7.

A.4 Random forest regression in R

Random forest regression can be implemented in R using a set of commands. The data for predictor (independent variables) and output variables are to be stored in the form of a data frame. In case of machine learning techniques the data are split into training and testing sets with an appropriate ratio. Consider a dataset with wind speed as output variable and its wavelet decomposition as predictor variables. As discussed in Chapter 5, the approximate and detail signals form the input dataset for predicting wind speed. In this example for RF model, we take a dataframe `f5` with 4320 samples and divide them into training (3000) and testing (1320) sets.

Example A.11. R codes for RF model

```
head(f5)
summary(f5)
library(caret)
library(caTools)
traindata<-f5[1:3000,]
testdata<-f5[3001:4320,]
xtrain<-traindata[,c(2,3,4,5,6,7)]
ytrain<-traindata[,1]
ytest<-testdata[,1]
xtest<-testdata[,c(2,3,4,5,6,7)]
library(randomForest)
rfrobj<-randomForest(ws~.,data=f5,
ntree=1000,keep.forest=TRUE, importance=TRUE,
na.action=na.omit)
summary(rfrobj)
predict.rfr<-predict(rfrobj,xtest)
write.table(predict.rfr, file = "rfr.txt", row.names = TRUE,
col.names = NA)
print(predict.rfr)
summary(predict.rfr)
rmse.rfr<-sqrt(mean((predict.rfr-ytest)^2))
nmse.rfr<-sum((predict.rfr-ytest)^2)
/sum((ytest-mean(ytest))^2)
r2.rfr<-sum((predict.rfr-mean(ytest))^2)
/sum((ytest-mean(ytest))^2)
plot(rfrobj)
lines(predict.rfr)
```

Further, to determine the optimal number of trees for the given dataset, R command `which.min(rfrobj$mse)` can be used. Fig. A.8 illustrates the MSE values with variation in number of trees.

A.5 Gradient boosted machines in R

Similar to random forest regression, gradient boost machines (GBMs) can be used to predict wind speed or any other variable. We now discuss the R codes for GBM model for the same example discussed in the previous section. See Fig. A.9.

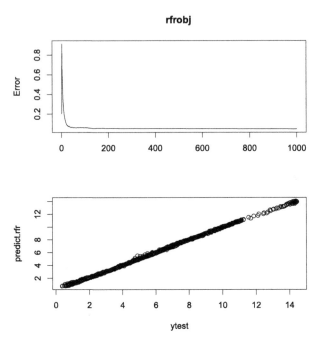

Figure A.8 Variation of MSE with number of trees and scatter plot between actual and predicted values for RF model.

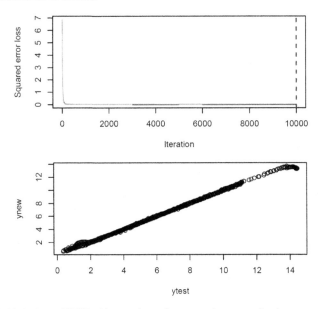

Figure A.9 Variation of MSE with number of trees and scatter plot between actual and predicted value for GBM model.

Example A.12. R codes for GBM

```
library(gbm)
library(xgboost)
library(h2o)
library(pdp)
library(ggplot2)
head(f5)
summary(f5) traindata<-ftc[1:3000,]
testdata<-ftc[3001:4320,]
xtrain<-traindata[,c(2,3,4,5,6,7)]
ytrain<-traindata[,1]
ytest<-testdata[,1]
xtest<-testdata[,c(2,3,4,5,6,7)]
gbm.fit<-gbm(formula=ws~.,data = f5,
var.monotone = c(0,0,0,0,0,0),
distribution = "gaussian", n.trees = 10000, shrinkage = 0.05,
interaction.depth = 3, bag.fraction = 0.5, train.fraction = 0.5,
n.minobsinnode = 10, cv.folds = 5, keep.data = TRUE,
verbose = FALSE, n.cores = 1)
print(gbm.fit)
sqrt(min(gbm.fit$cv.error))
gbm.perf(gbm.fit,method="cv")
ynew<-predict(gbm.fit, xtest, 9982, type = "link",
single.tree = FALSE)
write.table(ynew, file = "yGBM.txt", row.names = TRUE,
col.names = NA)
library(forecast)
accuracy(ynew,ytest) rmse.gbm<-sqrt(mean((ynew-ytest)^2))
nmse.gbm<-sum((ynew-ytest)^2)/sum((ytest-mean(ytest))^2)
r2.gbm<-sum((ynew-mean(ytest))^2)
/sum((ytest-mean(ytest))^2)
write.table(ytest, file = "ytest.txt" row.names = TRUE,
col.names = NA)
```

Index

Printed in the United States
By Bookmasters